3D-Computation of Incompressible Internal Flows

Edited by
Gabriel Sottas and
Inge L. Ryhming

Notes on Numerical Fluid Mechanics (NNFM) Volume 39

Volume 8 Vectorization of Computer Programs with Applications to Computational Fluid Dynamics (W. Gentzsch)

Volume 12 The Efficient Use of Vector Computers will Emphasis on Computational Fluid Dynamics (W. Schönauer / W. Gentzsch, Eds.)

Volume 13 Proceedings of the Sixth GAMM-Conference on Numerical Methods in Fluid Mechanics (D. Rues / W. Kordulla, Eds.)

Volume 14 Finite Approximations in Fluid Mechanics (E. H. Hirschel, Ed.)

Volume 17 Research in Numerical Fluid Dynamics (P. Wesseling, Ed.)

Volume 18 Numerical Simulation of Compressible Navier-Stokes Flows (M. W. Bristeau / R. Glowinski / J. Periaux / H. Viviand, Eds.)

Volume 20 Proceedings of the Seventh GAMM-Conference on Numerical Methods in Fluid Mechanics (M. Deville, Ed.)

Volume 22 Numerical Simulation of the Transonic DFVLR-F5 Wing Experiment (W. Kordulla, Ed.)

Volume 23 Robust Multi-Grid Methods (W. Hackbusch, Ed.)

Volume 26 Numerical Solution of Compressible Euler Flows (A. Dervieux / B. van Leer / J. Periaux / A. Rizzi, Eds.)

Volume 27 Numerical Simulation of Oscillatory Convection in Low-Pr Fluids (B. Roux, Ed.)

Volume 28 Vortical Solution of the Conical Euler Equations (K. G. Powell)

Volume 29 Proceedings of the Eighth GAMM-Conference on Numerical Methods in Fluid Mechanics (P. Wesseling, Ed.)

Volume 30 Numerical Treatment of the Navier-Stokes Equations (W. Hackbusch / R. Rannacher, Eds.)

Volume 31 Parallel Algorithms for Partial Differential Equations (W. Hackbusch, Ed.)

Volume 32 Adaptive Finite Element Solution Algorithm for the Euler Equations (R. A. Shapiro)

Volume 33 Numerical Techniques for Boundary Element Methods (W. Hackbusch, Ed.)

Volume 34 Numerical Solutions of the Euler Equations for Steady Flow Problems (A. Eberle / A. Rizzi / H. E. Hirschel)

Volume 35 Proceedings of the Ninth GAMM-Conference on Numerical Methods in Fluid Mechanics (J. B. Vos / A. Rizzi / I. L. Ryhming, Eds.)

Volume 36 Numerical Simulation of 3-D Incompressible Unsteady Viscous Laminar Flows (M. Deville / T.-H. Lê / Y. Morchoisne, Eds.)

Volume 37 Supercomputers and Their Performance in Computational Fluid Mechanics (K. Fujii, Ed.)

The addresses of the Editors and further titles of the series are listed at the end of the book.

3D-Computation of Incompressible Internal Flows

Proceedings of the GAMM Workshop
held at EPFL, 13–15 September 1989,
Lausanne, Switzerland

Edited by
Gabriel Sottas and
Inge L. Ryhming

Vieweg ist a subsidiary company of the Bertelsmann Publishing Group International.

Produced by W. Langelüddecke, Braunschweig
Printed on acid-free paper
Printed in Germany

ISSN 0179-9614
ISBN 3-531-07639-9

PREFACE

The aim of the 1989 GAMM Workshop on 3D-Computation of Incompressible Internal Flows was the simulation of a realistic incompressible flow field in an important industrial application. In view of the difficulties involved in formulating such a test case, requiring the availability of an experimental data base, extreme care had to be taken in the selection of the proper one.

Professor I.L. Ryhming's proposal, that the flow through a Francis turbine configuration or parts thereof would be feasible as a test case, because of the numerical challenges as well as the possibility to produce an experimental data base by using the experimental facilities of the *Hydraulic Machines and Fluid Mechanics Institute* (IMHEF) at the *Swiss Federal Institute of Technology* in Lausanne (EPFL), was accepted by the GAMM Committee in April 1987. A scientific committee, formed under the chairmanship of Professor I.L. Ryhming, met a few times to decide on the Francis turbine configuration, the test case specifications, etc., whereby the design input came from the water turbine experts. This committee decided to restrict the studies to the three following typical applications for the best operating point of the turbine :

- simulation of the 3D flow in a Francis runner in rotation
- simulation of the 3D flow in the distributor (stay and guide vane rings) of this turbine
- simulation of the 3D flow in an elbow draft tube

The simultaneous computation of two or three of these geometries was encouraged.

Professor P. Henry, and his experimentalist group at IMHEF, had agreed to design the experimental set up and to generate the experimental data base presented during the Workshop. A few surprises came up along the way. The most serious one was the flow behaviour in the simplified draft tube of the turbine, that did not allow for detailed measurements with the available instrumentation.

Nevertheless, we believe the workshop was successful : it was attended by over 40 scientists/engineers from 26 different industrial as well as academic groups representing 11 countries around the world. Among these attending groups, 14 presented flow simulation results. Thus, a total of 22 contributions were submitted : 2 simultaneous simulations of the flow in the distributor and the runner, 2 flow simulations in the distributor, 11 in the runner and 7 in the draft tube.

The publication of these proceedings has been delayed significantly due to several reasons. First of all, problems with proper definitions and inaccuracies in the experimental data base needed clarifications, and this absorbed much effort, time and, indeed, good will. Besides, the individual papers were reviewed by an ad-hoc editorial board, and it was necessary to send the papers back, sometimes several times, to have the authors meet general as well as workshop ground rules agreed upon beforehand. We want to thank here the authors for their patience and cooperation in working for this project's completion. Now, in the middle of 1992, we present these Proceedings, knowing that the state of the art in the theoretical as well as in the experimental context has improved from late 1989 and with the hope that it will continue to improve and gain from this workshop's experience in the years to come.

As editors of these proceedings and organizers of the workshop, we would like to thank all the participants and to acknowledge the support from :

- Ecole Polytechnique Fédérale de Lausanne
- GAMM
- CRAY Research
- Banque Cantonale Vaudoise
- Société de Banque Suisse
- Ville de Lausanne

Special thanks are also due to all the IMHEF members who helped us to organize successfully this workshop. Three among them deserves to be explicitely mentioned : Dr. A. Drotz for his commitment during the setting up stage, Dr. A. Bottaro for his invaluable help in preparing the synthesis of the contributed numerical results, and Mrs C. Berg who arranged all the practical details around the meeting in an efficient way.

Lausanne, July 1992.

I.L. Ryhming G. Sottas

CONTENTS

Part 1

TEST CASES SPECIFICATIONS

TEST CASES SPECIFICATIONS

INTRODUCTION

The main goal of the 1989 GAMM Workshop entitled "*3D-Computation of Incompressible Internal Flows*" was to bring together specialists interested in the calculation of flow related to hydraulic turbines. A Francis turbine geometry was decided upon and offered as a test case for the flow simulation.

The complex geometry of the turbine can be viewed as the assembly of four main hydraulic components : the spiral casing, the distributor, the runner and the draft tube (see Figure 1). For this workshop we restricted the studies to the last three components, namely the computation of the 3D flow in the distributor (stay vanes and wicket gate), the Francis runner (rotating part) and the elbow draft tube.

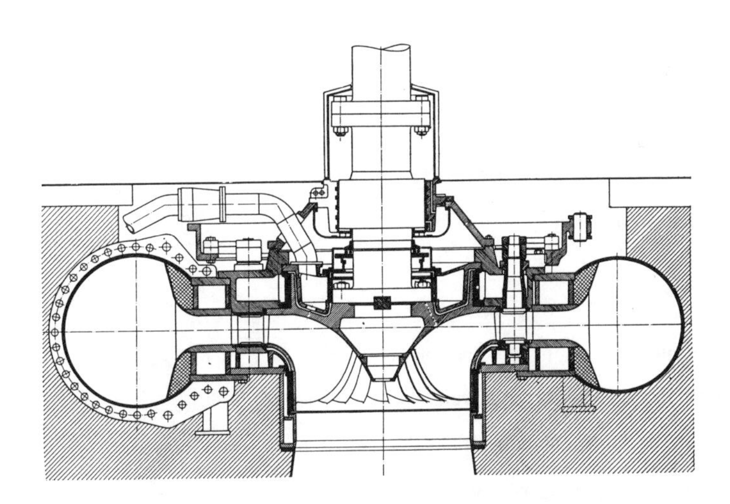

Fig. 1. Axial view of a Francis turbine

The test case specifications were designed in such a way that each contributor could choose to simulate separately the flow in each of these components. But, the possibility was also offered to compute the flow simultaneously in a combination of two of these components (e.g. by a multiblock approach) or in the entire turbine.

A model Francis turbine was built according to the specified geometry and tested in the IMHEF Universal Hydraulic Machine Test Facility. Thus, an experimental data base was built for the best efficiency operating point. This data base comprises integral properties of the flow, such as torque and global flow rate, detailed pressure and velocity distribution measurements on some specified (non moving) axes and pressure distribution measurements on the blades of the runner in motion.

A detailed presentation of the test case Francis turbine geometry and of the flow survey performed at IMHEF is given in the paper entitled *"Experimental Flow study of the GAMM turbine model"* in part 2 of these Proceedings. Accordingly, a description of the three test cases will be given below only in terms of the geometrical and physical data that were at the disposal of the contributors to perform their calculations. All these data are available from the editors.

NOMENCLATURE

C	:	absolute velocity norm		[m/s]
C_R	:	absolute radial velocity		[m/s]
C_θ	:	absolute tangential velocity		[m/s]
C_Z	:	absolute axial velocity		[m/s]
E	:	specific hydraulic energy		[J/kg]
P	:	static pressure		[N/m²]
P_{ref}	:	reference static pressure		[N/m²]
R_{ref}	:	reference radius		[m]
T	:	global runner torque		[Nm]
$\dot{V}$	:	flow rate		[m³/s]
c^*	:	normalized absolute velocity norm	$= C / \sqrt{2E}$	[-]
c_R^*	:	normalized absolute radial velocity	$= C_R / \sqrt{2E}$	[-]
c_θ^*	:	normalized absolute tangential velocity	$= C_\theta / \sqrt{2E}$	[-]
c_Z^*	:	normalized absolute axial velocity	$= C_Z / \sqrt{2E}$	[-]
c_m^*	:	normalized absolute meridional velocity	$= \sqrt{(c_R^*)^2 + (c_Z^*)^2}$	[-]
c_u^*	:	normalized absolute tangential velocity	$= c_\theta^*$	[-]
c_a^*	:	normalized absolute axial velocity	$= c_Z^*$	[-]
c_p^*	:	normalized pressure coefficient	$= (P - P_{ref}) / (\rho E)$	[-]
g	:	gravity constant		[m/s²]
t^*	:	normalized torque	$= T / (\rho \pi E R_{ref}^3)$	[-]
u^*	:	normalized peripheral velocity norm		[-]
w^*	:	normalized relative velocity norm		[-]

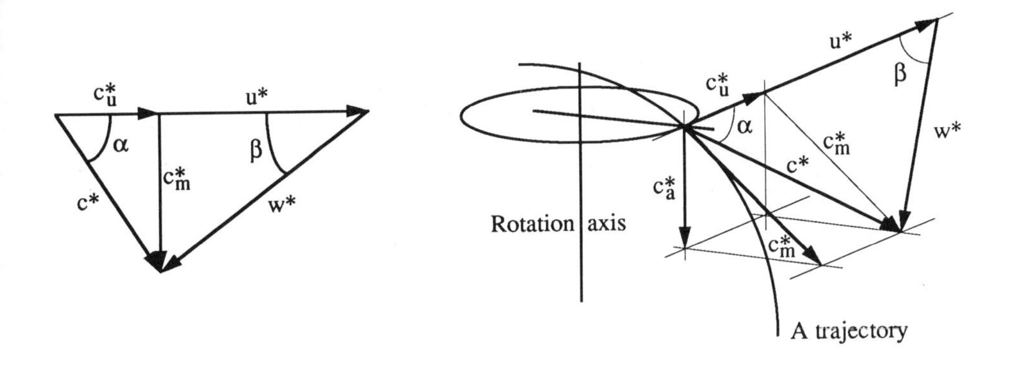

Fig. 2. Velocity components

α	: absolute flow angle	$= \text{arctg}(\frac{c_m^*}{c_u^*})$ for the runner	[°]
		$= \text{arccos}(\frac{c_\theta^*}{c^*})$	[°]
β	: relative flow angle	$= \text{arctg}(\frac{c_m^*}{(\omega R / \sqrt{2E}) - c_u^*})$	[°]
ρ	: water density		[kg/m³]
η	: efficiency	$= (T\omega) / (\rho \dot{V} E)$	[-]
ω	: angular velocity		[rad/s]
φ	: discharge coefficient	$= \dot{V} / (\pi \omega R_{ref}^3)$	[-]
ψ	: energy coefficient	$= 2E / (\omega^2 R_{ref}^2)$	[-]

COORDINATE SYSTEM (X,Y,Z)

To simplify the geometrical description of the whole Francis turbine, the same coordinate system (X,Y,Z) has been used for the three test cases. It has been fixed according to the three following conditions :

- the (X,Y) plane is placed at half the height of the distributor,
- the X axis is parallel to the draft tube symmetry plane,
- the Z axis coincides with the turbine axis and is directed from the lower to the upper ring of the distributor.

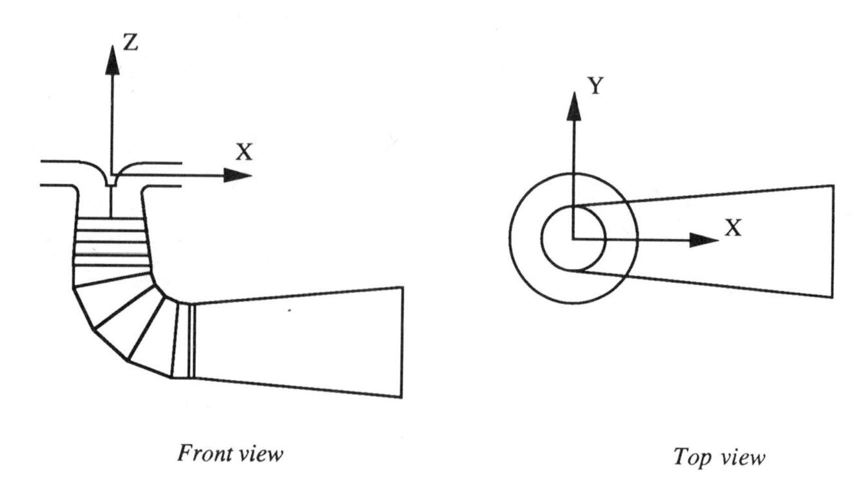

Fig. 3. Schematic Francis turbine with the coordinate system

TEST CASE 1 : DISTRIBUTOR

DISTRIBUTOR GEOMETRY

The following subsections give a detailed description of the data defining the distributor geometry.

Upper and lower ring

The upper and lower rings are specified as follows :

Upper ring : $\{ N_{pu}, \{R, Z\}_{N_{pu}} \}$

N_{pu}	number of upper ring points
$\{R, Z\}_{N_{pu}}$	polar coordinates of the upper ring points

Lower ring : $\{ N_{pl}, \{R, Z\}_{N_{pl}} \}$

N_{pl}	number of lower ring points
$\{R, Z\}_{N_{pl}}$	polar coordinates of the lower ring points

Upstream and downstream limits

The upstream and downstream limits are the revolution surfaces generated by the measurement axis. The inlet axis is vertical and intersect the lower ring at R = 346.0 [mm]

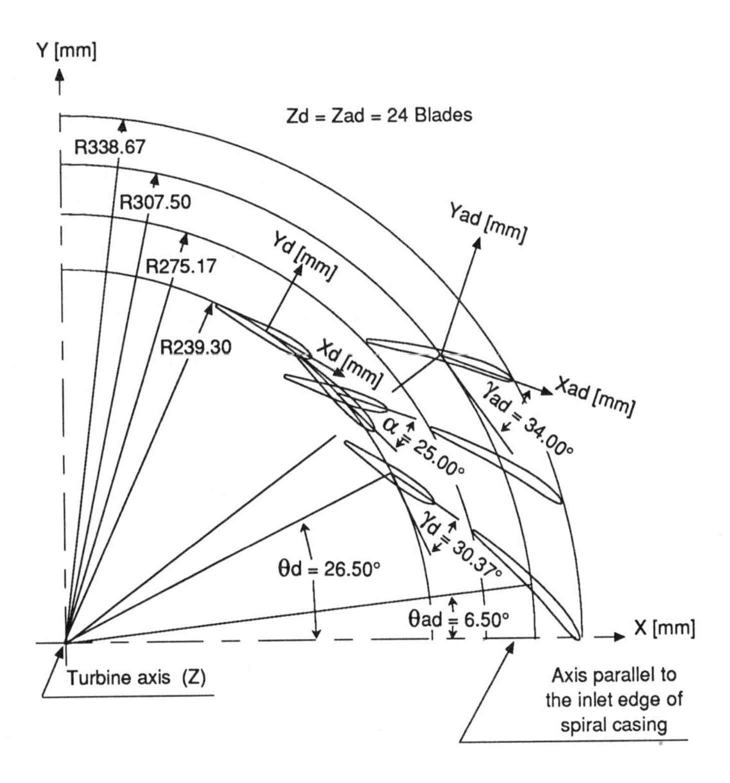

Fig. 4. Distributor geometry

and $Z = -60.0$ [mm] whereas the outlet one is oblique, making an angle of $20°$ with the vertical, and intersect the lower ring at $R = 210.9$ [mm] and $Z = -62.78$ [mm]. We observe that this outlet surface also constitutes the inlet surface for the runner geometry.

Blades

The blades are vertical. Consequently, they are specified by simply giving one single section in the (X,Y) plane. The geometrical data of the blades contain the leading and the trailing edge points. Moreover, we observe that the blades have thick trailing edges.

The data for the blade sections are given in the following sense : trailing edge towards leading edge on the suction side, leading edge towards trailing edge on the pressure side.

The blades are specified in the following way :

a) by the sections :

Stay vane : $\{ N_{ad}, R_{ad}, \theta_{ad}, \delta\theta_{ad}, \gamma_{ad}, \{X, Y\}_{N_{ad}} \}$

N_{ad} number of points per sections

$R_{ad}, \theta_{ad}, \delta\theta_{ad}, \gamma_{ad}$	radius and angles defining the position of the blade in the (X,Y,Z) coordinate system (see Figure 4, $\delta\theta_{ad}$ is the angle between two consecutive blades)
$\{X, Y\}_{N_{ad}}$	cartesian coordinates of the blade contour in the local coordinate system (see Figure 4)

Wicket gate : $\{ N_d, R_d, \theta_d, \delta\theta_d, \gamma_d, \{X, Y\}_{N_d} \}$

N_d	number of points per sections
$R_d, \theta_d, \delta\theta_d, \gamma_d$	radius and angles defining the position of the blade in the (X,Y,Z) coordinate system (see Figure 4, $\delta\theta_d$ is the angle between two consecutive blades)
$\{X, Y\}_{N_d}$	cartesian coordinates of the blade contour in the local coordinate system (see Figure 4)

b) by the camber lines :

Stay vane : $\{ N_{ad}, \{R, \theta, Z, n_R, n_\theta, n_Z, Wp\}_{N_{ad}} \}$

N_{ad}	number of camber line points
$\{R, \theta, Z\}_{N_{ad}}$	cylindrical coordinates of the camber line points
$\{n_R, n_\theta, n_Z\}_{N_{ad}}$	normal at every camber line point (in the cylindrical referential)
$\{Wp\}_{N_{ad}}$	width of the profile at every camber line point

Wicket gate : $\{ N_d, \{R, \theta, Z, n_R, n_\theta, n_Z, Wp\}_{N_d} \}$

N_d	number of camber line points

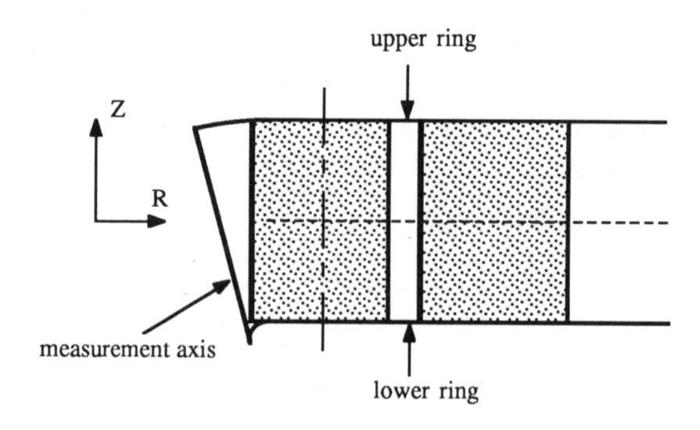

Fig. 5. Meridional view of the distributor

$\{R, \theta, Z\}_{N_d}$	cylindrical coordinates of the camber line points
$\{n_R, n_\theta, n_Z\}_{N_d}$	normal at every camber line point (in the cylindrical referential)
$\{Wp\}_{N_d}$	width of the profile at every camber line point

PHYSICAL DATA PROVIDED FOR THE FLOW SIMULATION IN THE DISTRIBUTOR

The following data are given :

- the specific hydraulic energy between ref. sections I & $\bar{2}$ $E_{I\div\bar{2}}$ = 58.42 [J/kg]
- the reference radius R_{ref} = 0.200 [m]
- the reference static pressure P_{ref} = 94300.000 [N/m²]
- the gravity constant g = 9.806 [m/s²]
- the flow rate $\dot{V}$ = 0.372 [m³/s]
- the water density ρ = 1000.000 [kg/m³]
- the pressure distribution $P - P_{ref}$ measured at the inlet and at the outlet of the distributor
- the velocity distribution (C_R, C_θ, C_Z) measured at the inlet and at the outlet of the distributor

The pressure distribution $P - P_{ref}$ and the velocity distribution (C_R, C_θ, C_Z) measured at the inlet and at the outlet of the distributor, together with the location of the measurement points, are specified in the following way :

Inlet data : $\{ N_a, N_{pa}, \{ \{R, \theta, Z, P - P_{ref}, C_R, C_\theta, C_Z\}_{N_{pa}} \}_{N_a} \}$

Outlet data : $\{ N_a, N_{pa}, \{ \{R, \theta, Z, P - P_{ref}, C_R, C_\theta, C_Z\}_{N_{pa}} \}_{N_a} \}$

N_a	number of measurement axis
N_{pa}	number of points per measurement axis
$\{ \{R, \theta, Z\}_{N_{pa}} \}_{N_a}$	cylindrical coordinates of the measurement points
$\{ \{P - P_{ref}\}_{N_{pa}} \}_{N_a}$	pressure at each measurement points
$\{ \{C_R, C_\theta, C_Z\}_{N_{pa}} \}_{N_a}$	velocity components at each measurement points

These data, in their normalized form are presented in Figure 6 for the inlet, and in Figure 7 for the outlet axis, respectively.

For the inlet axis (stayring inlet), three sets of measurements are at disposal, corresponding to three different angular positions of the probe axis (namely $\theta = 265$ [°], 355 [°] and 175 [°]). For comparison purposes during the workshop, contributors willing to use experimental data to specify the inlet boundary conditions have been asked to use those corresponding to $\theta = 265$ [°].

Distributor - Inlet axis

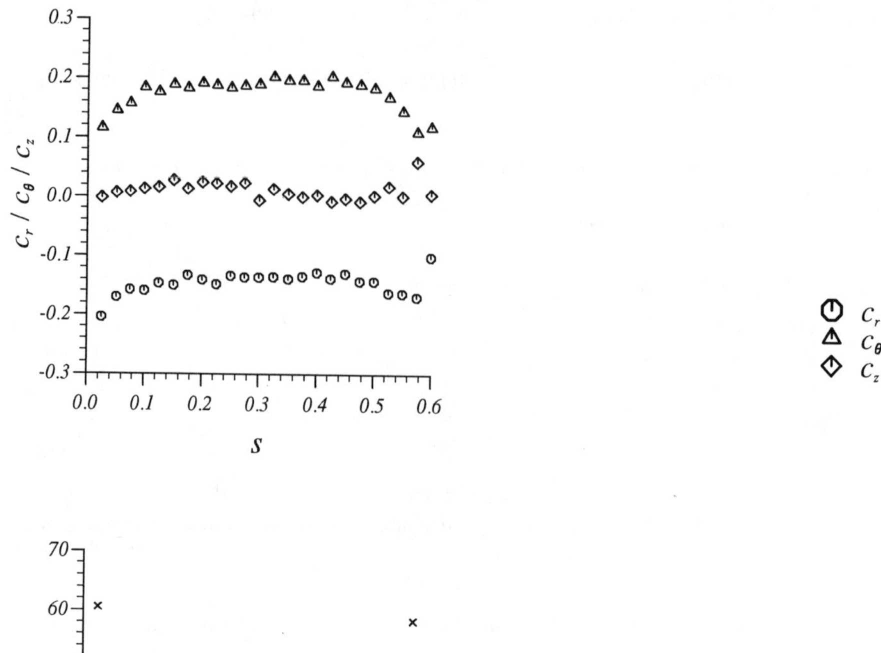

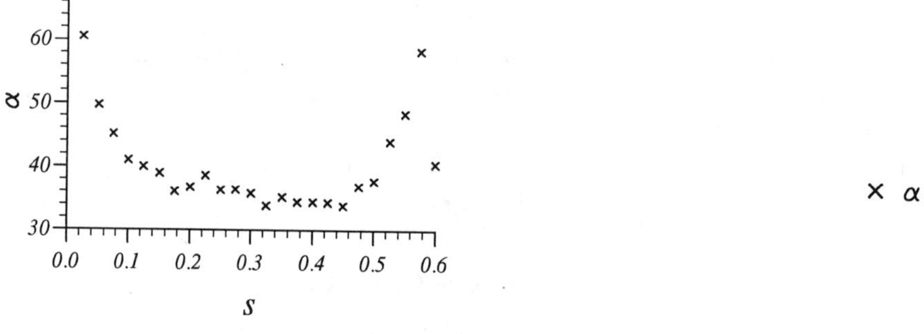

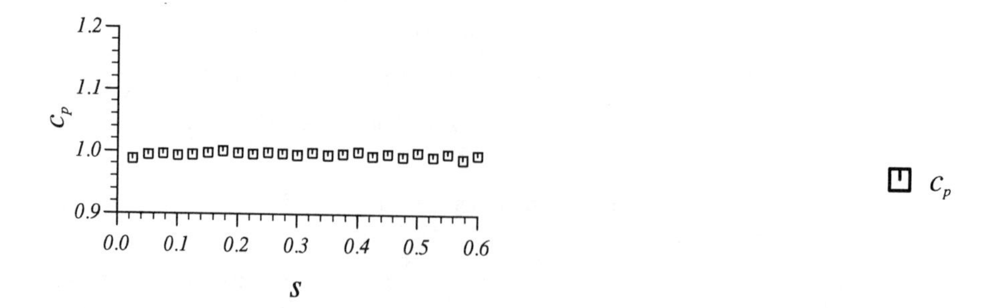

Fig. 6. Experimental velocity components, flow angle and pressure coefficient at the distributor inlet axis.

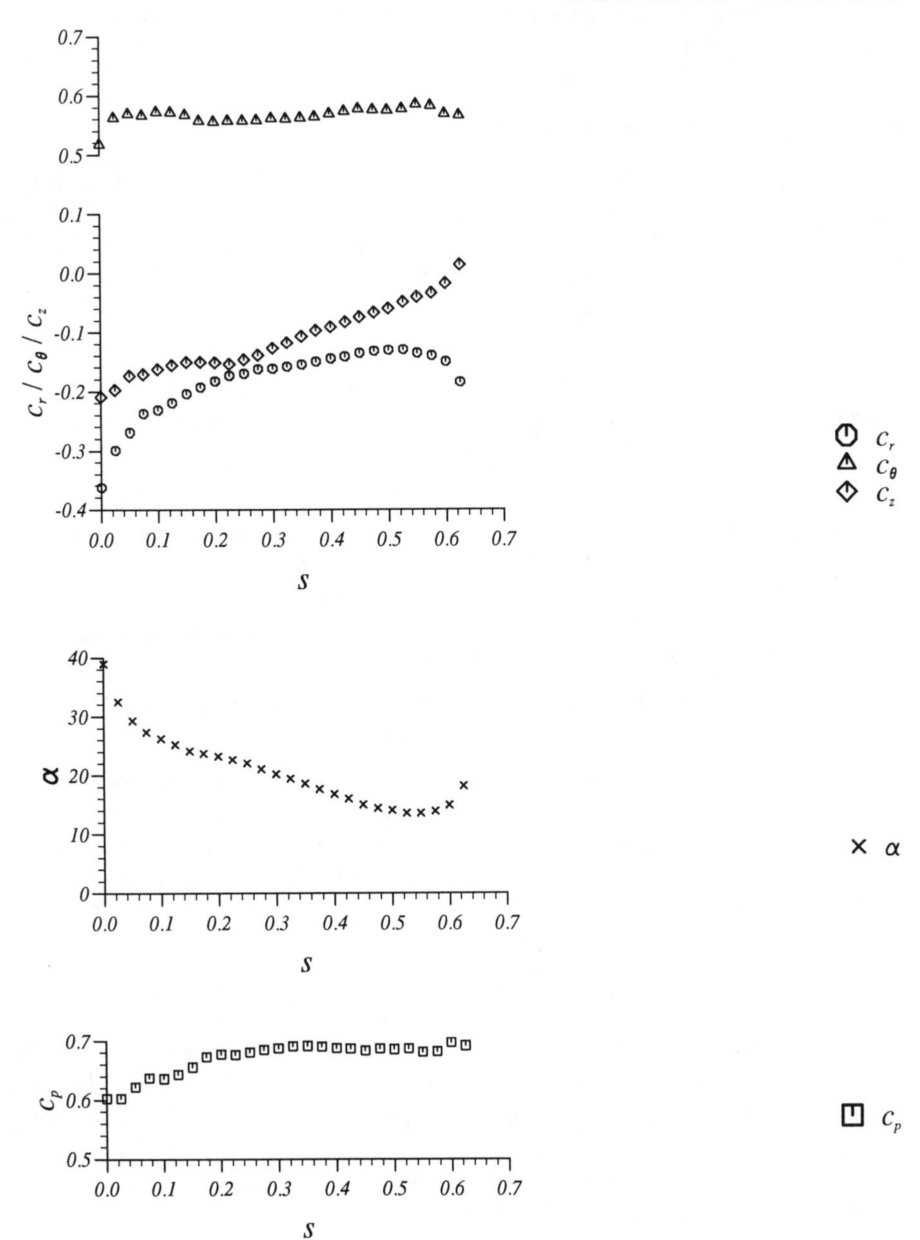

Fig. 7. Experimental velocity components, flow angle and pressure coefficient at the distributor outlet axis.

At the outlet axis, the mean flow is seen as axisymmetric by the pressure probe. Thus only one set of data is given, corresponding to the overall average of the measurements carried out at each probe passage.

REQUIRED RESULTS FOR THE DISTRIBUTOR FLOW

We recall here that when specified by the camber lines, the description of the geometry of the stay vane blades and of the wicket gate blades contains the leading and the trailing edge point (and that the width of the profiles at the trailing edge is different from zero !). This fact is used in the description of the required results given below.

Requirement 1

c_p^* along three sections at constant height (Z = –50, 0, 50 [mm]) of the stay vane and of the wicket gate blades versus the normalized arclength along the section's camber line.

$\{ \{ I, N_I, \{ Z, \{R, \theta, s^*, (c_p^*)_i, (c_p^*)_e\}_{N_I} \}_Z \}_I \}$

I	indicator which value belongs to the set { *stay_vane, wicket_gate* }
N_I	number of camber line points
Z	height of the chosen cuts of the blades (Z = *–50, 0 or 50 [mm]*)
$\{ \{ \{R, \theta\}_{N_I} \}_Z \}_I$	cylindrical coordinates of the camber line points per sections
$\{ \{ \{s^*\}_{N_I} \}_Z \}_I$	normalized arclength corresponding to every camber line point
$\{ \{ \{(c_p^*)_i, (c_p^*)_e\}_{N_I} \}_Z \}_I$	normalized pressure coefficient at the blade points corresponding to every camber line point (one on the pressure side and one on the suction side of the blade)

Note that : N_I and $\{ \{ \{R, \theta\}_{N_I} \}_Z \}_I$ are part of the given geometrical data.

Let one of the above sections be chosen (I = stay_vane or wicket_gate and Z = –50, 0 or 50) and let (R, θ) be one point of the camber line of this section. The arclength s associated with the point (R, θ) is the length of the portion of camber line joining the leading edge point of this camber line to the point (R, θ) and is computed assuming the camber line is piecewise linear (constituted of $(N_I - 1)$ segments). The associated normalized arclength s^* is then simply the arclength s divided by R_{ref}.

Two blade points are naturally associated with the camber line point (R, θ), one on the pressure side and one on the suction side of the blade (they are constructed by using the normal vector and the width of the profile associated with the camber line point). $(c_p^*)_i$ and $(c_p^*)_e$ are the normalized pressure coefficients evaluated at these two points.

Requirement 2

c_p^*, c_R^*, c_θ^*, c_Z^* and α along the distributor inlet measurement axis versus the normalized arclength along the measurement axis.

$\{ N_{pa}, \{R, Z, s^*, c_p^*, c_R^*, c_\theta^*, c_Z^*, \alpha\}_{N_{pa}} \}$

N_{pa}	number of points per measurement axis
$\{R, Z\}_{N_{pa}}$	cylindrical coordinates of the measurement points
$\{s^*\}_{N_{pa}}$	normalized arclength corresponding to every measurement point
$\{ c_p^* \}_{N_{pa}}$	normalized pressure coefficient at each measurement point
$\{ c_R^*, c_\theta^*, c_Z^* \}_{N_{pa}}$	normalized velocity components at each measurement point
$\{\alpha\}_{N_{pa}}$	absolute flow angle

Note that : N_{pa} and $\{R, Z\}_{N_{pa}}$ are part of the given measured data.

In order to fix the axis position we choose $\theta_{inlet} = 265.0000$ [°] (i.e. we choose the measurement axis 1 among the three data sets).

The arclength s associated with a point (R, Z) on the axis is the length of the portion of axis joining the lower ring point of the axis to the point (R, Z). The associated normalized arclength s^* is then simply the arclength s divided by R_{ref}. We recall here (see the geometrical description) that the intersection of the inlet axis with the lower ring is at R = 346.0 [mm] and Z = –60.0 [mm].

Requirement 3

averaged c_p^*, c_R^*, c_θ^*, c_Z^* and α along the distributor outlet measurement axis versus the normalized arclength along the measurement axis.

$\{ N_{pa}, \{R, Z, s^*, (c_p^*)_{av}, (c_R^*)_{av}, (c_\theta^*)_{av}, (c_Z^*)_{av}, \alpha\}_{N_{pa}} \}$

N_{pa}	number of points per measurement axis
$\{R, Z\}_{N_{pa}}$	cylindrical coordinates of the measurement points
$\{s^*\}_{N_{pa}}$	normalized arclength corresponding to every measurement point
$\{ (c_p^*)_{av} \}_{N_{pa}}$	averaged normalized pressure coefficient at each measurement point
$\{(c_R^*)_{av},(c_\theta^*)_{av},(c_Z^*)_{av}\}_{N_{pa}}$	averaged normalized velocity components at each measurement point

$\{\alpha\}_{N_{pa}}$	absolute flow angle (corresponding to the averaged velocity)

Note that : N_{pa} and $\{R, Z\}_{N_{pa}}$ are part of the given measured data.

The arclength s associated with a point (R, Z) on the axis is the length of the portion of axis joining the lower ring point of the axis to the point (R, Z). The associated normalized arclength s^* is then simply the arclength s divided by R_{ref}. We recall here (see the geometrical description) that the intersection of the outlet axis with the lower ring is at R = 210.9 [mm] and Z = –62.78 [mm].

At the outlet section, the hypothesis of axisymmetry is made. Thus the computed values c_p^*, c_R^*, c_θ^*, c_Z^* should be averaged on circles of radius R and located at constant height Z.

Requirement 4

c_p^*, c_R^*, c_θ^*, c_Z^* and α along the leading and the trailing edges of the stay vane blade and of the wicket gate blade versus the normalized arclength along the edge.

$$\{\ N_e,\ \{\ I,\ \{I_e,\ \{R, \theta, Z, s^*, c_p^*, c_R^*, c_\theta^*, c_Z^*, \alpha\}_{N_e}\ \}_{I_e}\ \}_I\ \}$$

N_e	number of points on each edges
I	indicator which value belongs to the set { *stay_vane*, *wicket_gate* }
I_e	indicator of the edge of the blade (which value belongs to the set { *leading*, *trailing* })
$\{\ \{\ \{R, \theta, Z\}_{N_e}\ \}_{I_e}\ \}_I$	cylindrical coordinates of the blade edge points
$\{\ \{\ \{s^*\}_{N_e}\ \}_{I_e}\ \}_I$	normalized arclength corresponding to every blade edge point
$\{\ \{\ \{\ c_p^*\ \}_{N_e}\ \}_{I_e}\ \}_I$	normalized pressure coefficient at each blade edge point
$\{\ \{\ \{\ c_R^*, c_\theta^*, c_Z^*\ \}_{N_e}\ \}_{I_e}\ \}_I$	normalized velocity components at each blade edge point
$\{\ \{\ \{\alpha\}_{N_e}\ \}_{I_e}\ \}_I$	absolute flow angle corresponding to the velocity components

Note that : $\{\ \{\ \{R, \theta\}_{N_e}\ \}_{I_e}\ \}_I$ is part of the given geometrical data.

For each edge, the quantities have to be evaluated at points having a normalized arclength equal to $n*0.025$ where n is an integer between 0 and 24 (i.e. $N_e = 25$).

Let one of the edges be chosen (I = stay_vane or wicket_gate and I_e = leading or trailing) and let (R, θ, Z) be one of the given points on it. The arclength s associated with the point (R, θ, Z) is the length of the portion of the edge joining the point (R, θ, Z) to the point belonging to the edge and located at the height Z = –60 [mm] (which is the intersection of the edge with the lower ring except for the trailing edge of the wicket gate, see Figure 5). The associated normalized arclength s^* is then simply the arclength s divided by R_{ref}.

Requirement 5

computer related quantities :

- computer type,
- number of grid points used,
- cpu time,
- (cpu time) / iteration,
- (cpu time) / (number of grid cells),
- (cpu time) / (iteration * number of grid cells),
- ratio of vectorisation : cpu time (scalar version) / cpu time (vectorised version),
- parallelisation speedup,
- convergence criterion and convergence diagrams (if an iterative approach is used).

TEST CASE 2 : FRANCIS RUNNER

RUNNER GEOMETRY

The following subsections give a detailed description of the data defining the runner geometry.

Meridional channel

From the upstream and towards the downstream, the crown and the band of the runner are specified as follows :

Crown :	$\{ N_{pc}, \{R, Z\}_{N_{pc}} \}$	
	N_{pc}	number of crown points
	$\{R, Z\}_{N_{pc}}$	polar coordinates of the crown points
Band :	$\{ N_{pb}, \{R, Z\}_{N_{pb}} \}$	
	N_{pb}	number of band points
	$\{R, Z\}_{N_{pb}}$	polar coordinates of the band points

Upstream and downstream limits

The upstream limit is the revolution surface generated by the measurement axis. The inlet axis is oblique, making an angle of 20° with the vertical, and intersect the band at R = 210.9 [mm] and Z = –62.78 [mm]. We observe that this inlet surface also constitutes the outlet surface for the distributor geometry.

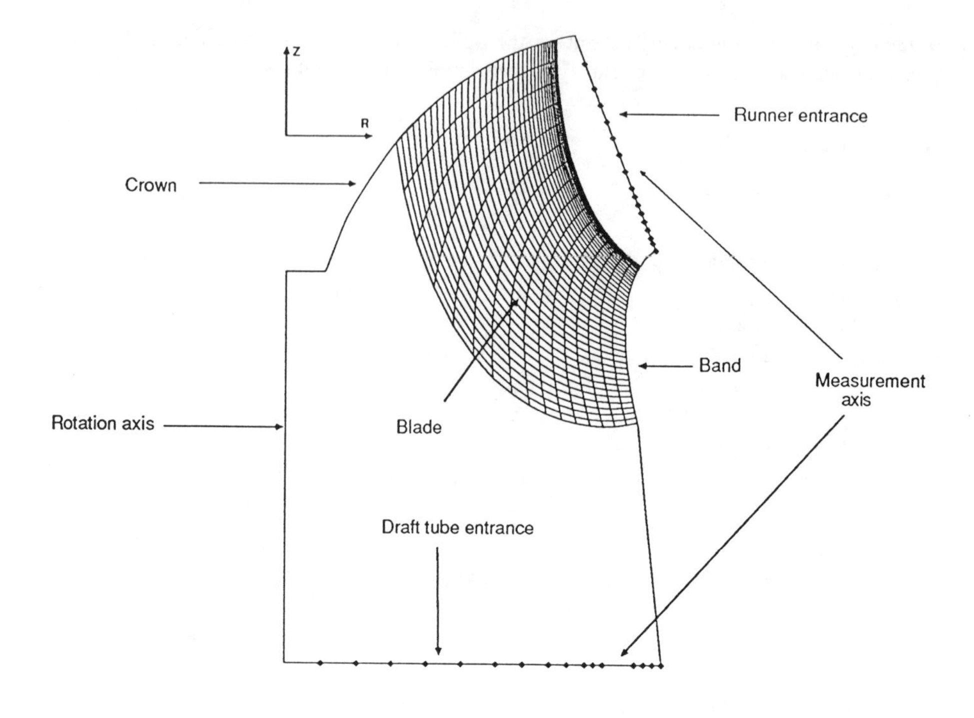

Figure 8. *Meridional view of the runner*

The downstream limit is the horizontal surface generated by the measurement axis. The outlet axis is horizontal and intersect the band at $R = 218.38$ [mm] and $Z = -346.35$ [mm]. We observe that this outlet surface also constitutes the inlet surface for the draft tube geometry.

Blades

The blades are specified using N_s (= 17) sections. These sections are theoretical stream surfaces of revolution. Their trailing edge is in a radial plane (X,Z) and is thin. The geometrical data of the blades contain the leading and the trailing edge points.

The data for the blade sections are given in the following sense : trailing edge towards leading edge on the suction side, leading edge towards trailing edge on the pressure side.

The blades are specified in the following way :

a) by the sections :

Blade :	$\{ \delta\theta, N_s, N_{ps}, \{ \{X, Y, Z, R, \theta\}_{N_{ps}} \}_{N_s} \}$	
	$\delta\theta = 2\pi/13$	periodicity angle
	N_s	number of sections
	N_{ps}	number of points per sections
	$\{ \{X, Y, Z\}_{N_{ps}} \}_{N_s}$	cartesian coordinates of the blade section points
	$\{ \{R, \theta, Z\}_{N_{ps}} \}_{N_s}$	cylindrical coordinates of the blade section points

b) by the camber lines :

Blade :	$\{ \delta\theta, N_s, N_{ps}, \{ \{R, \theta, Z, n_R, n_\theta, n_Z, Wp\}_{N_{ps}} \}_{N_s} \}$	
	$\delta\theta = 2\pi/13$	periodicity angle
	N_s	number of sections
	N_{ps}	number of camber line points per sections
	$\{ \{R, \theta, Z\}_{N_{ps}} \}_{N_s}$	cylindrical coordinates of the camber line points per sections
	$\{ \{n_R, n_\theta, n_Z\}_{N_{ps}} \}_{N_s}$	normal at every camber line point (in the cylindrical referential)
	$\{ \{Wp\}_{N_{ps}} \}_{N_s}$	width of the profile at every camber line point

PHYSICAL DATA PROVIDED FOR THE FLOW SIMULATION IN THE RUNNER

The following data are given :

• the specific hydraulic energy between ref. sections I & $\overline{2}$	$E_{I \div \overline{2}}$	=	58.42	[J/kg]
• the reference radius	R_{ref}	=	0.200	[m]
• the reference static pressure	P_{ref}	=	94300.000	[N/m^2]
• the gravity constant	g	=	9.806	[m/s^2]
• the global runner torque	T	=	375.54	[Nm]
• the flow rate	$\dot{V}$	=	0.372	[m^3/s]
• the water density	ρ	=	1000.000	[kg/m^3]
• the angular velocity	ω	=	52.36	[rad/s]

- the pressure distribution $P - P_{ref}$ measured at the inlet and at the outlet of the runner
- the velocity distribution (C_R, C_θ, C_Z) measured at the inlet and at the outlet of the runner

The pressure distribution $P - P_{ref}$ and the velocity distribution (C_R, C_θ, C_Z) measured at the inlet and at the outlet of the runner, together with the location of the measurement points, are specified in the following way :

Inlet data : $\{ N_a, N_{pa}, \{ \{R, \theta, Z, P - P_{ref}, C_R, C_\theta, C_Z\}_{N_{pa}} \}_{N_a} \}$

Outlet data : $\{ N_a, N_{pa}, \{ \{R, \theta, Z, P - P_{ref}, C_R, C_\theta, C_Z\}_{N_{pa}} \}_{N_a} \}$

N_a	number of measurement axis
N_{pa}	number of points per measurement axis
$\{ \{R, \theta, Z\}_{N_{pa}} \}_{N_a}$	cylindrical coordinates of the measurement points
$\{ \{P - P_{ref}\}_{N_{pa}} \}_{N_a}$	pressure at each measurement points
$\{ \{C_R, C_\theta, C_Z\}_{N_{pa}} \}_{N_a}$	velocity components at each measurement points

These data, in their normalized form are presented in Figure 9 for the inlet, and in Figure 10 for the outlet axis, respectively.

At the inlet axis, the mean flow is seen as axisymmetric by the pressure probe. Thus the given set of data corresponds to the overall average of the measurements carried out at each probe passage.

Since the outlet axis do not rotate with the runner, the measured data also are averaged values.

RUNNER FLOW DATA AVAILABLE ONLY AT THE WORKSHOP TIME

Some additional experimental data became available at the workshop time. These are :

- The pressure distribution c_p^* on the blade sections 2, 9 and 15. Some of the measurement points are located on the pressure side of the blades and some others are distributed on the suction side. These data are presented in Figure 11.
- The velocity distribution c_a^*, c_u^*, c_m^* measured on the runner middle axis. This "middle" axis is an oblique axis located just under the trailing edges of the blades. It makes an angle of 55° with the vertical, and intersect the band at $R = 205.09$ [mm] and $Z = -208.47$ [mm]. Since this axis do not rotate with the runner, the measured data are averaged values. These data are presented in Figure 12.

REQUIRED RESULTS FOR THE RUNNER FLOW

We recall here that the geometry of the blades has been defined using N_s (= 17) sections. Moreover, when specified by the camber lines, the description of each section contains the leading and the trailing edge points. These two facts are used in the description of the required results given below.

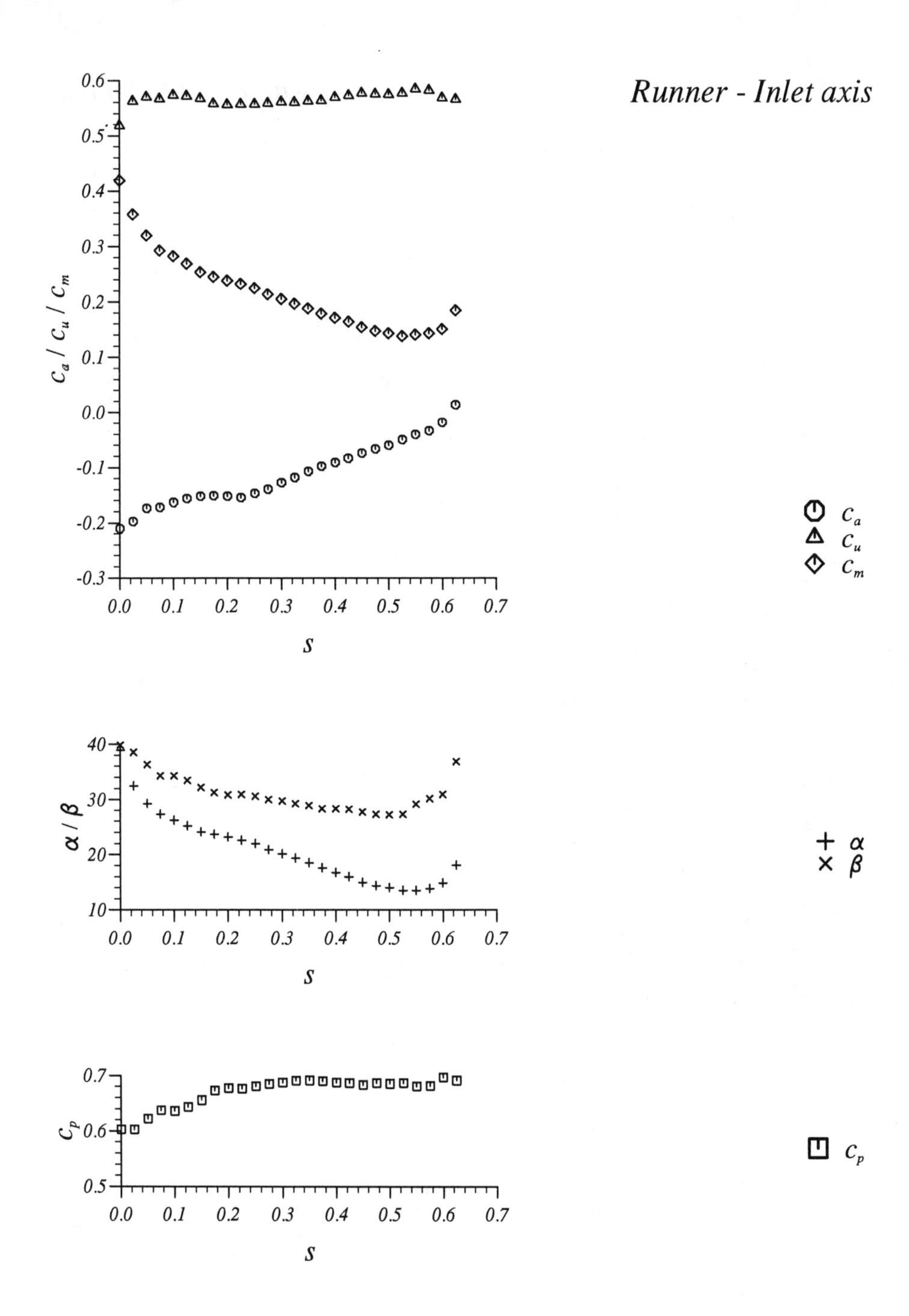

Fig. 9. Experimental velocity components, flow angles and pressure coefficient at the runner inlet axis.

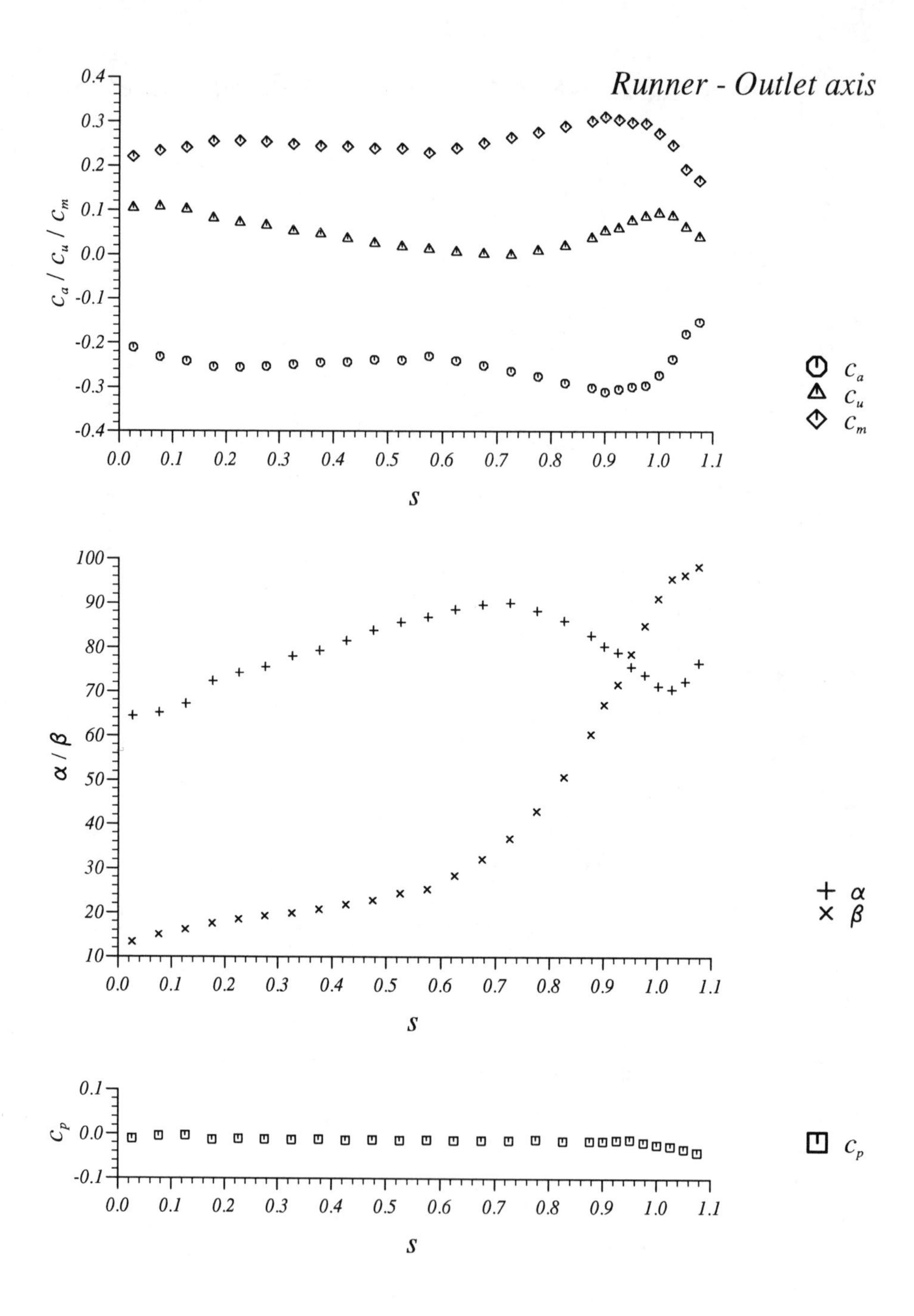

Fig. 10. Experimental velocity components, flow angles and pressure coefficient at the runner outlet axis.

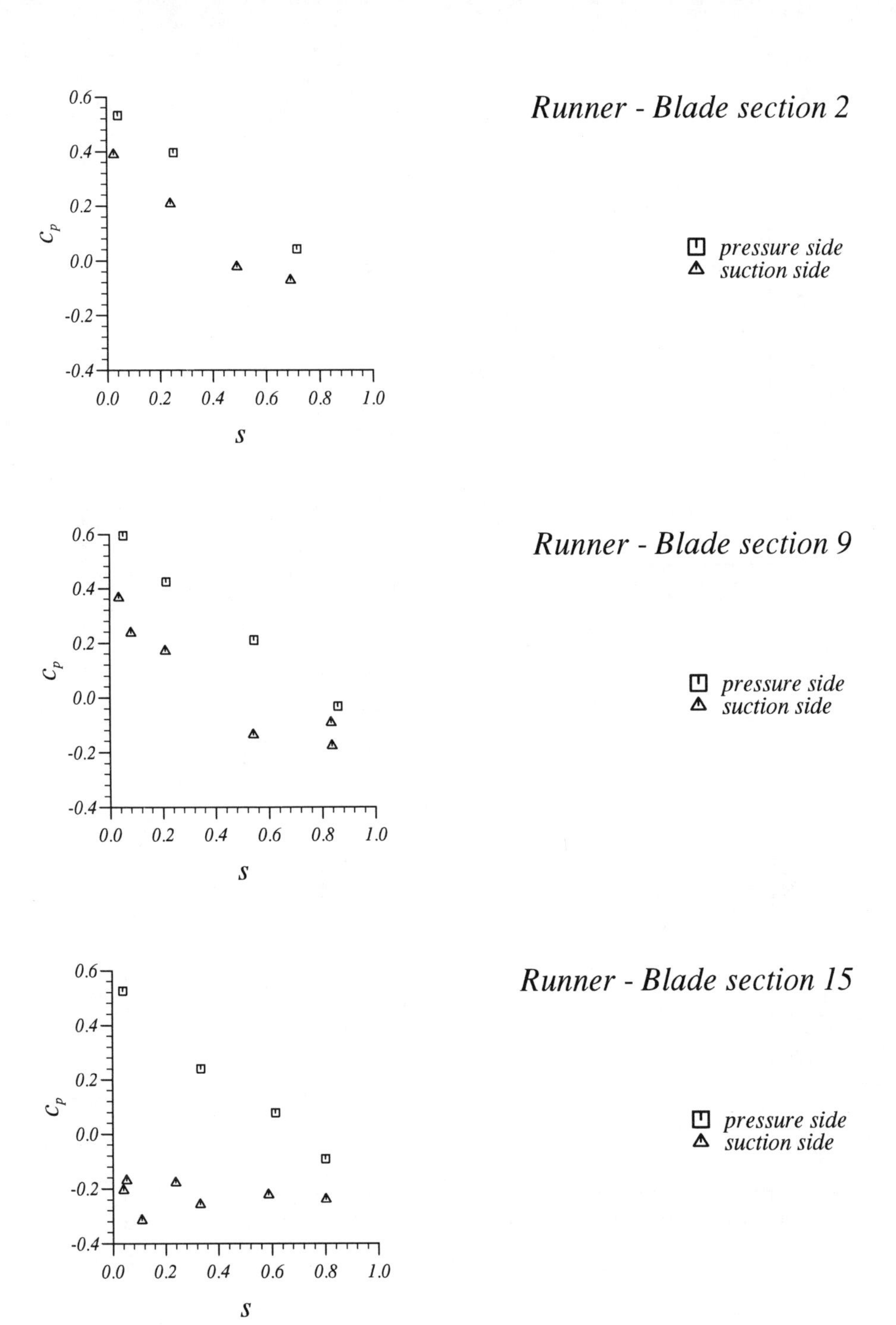

Fig. 11. Experimental pressure coefficient on the runner blade sections.

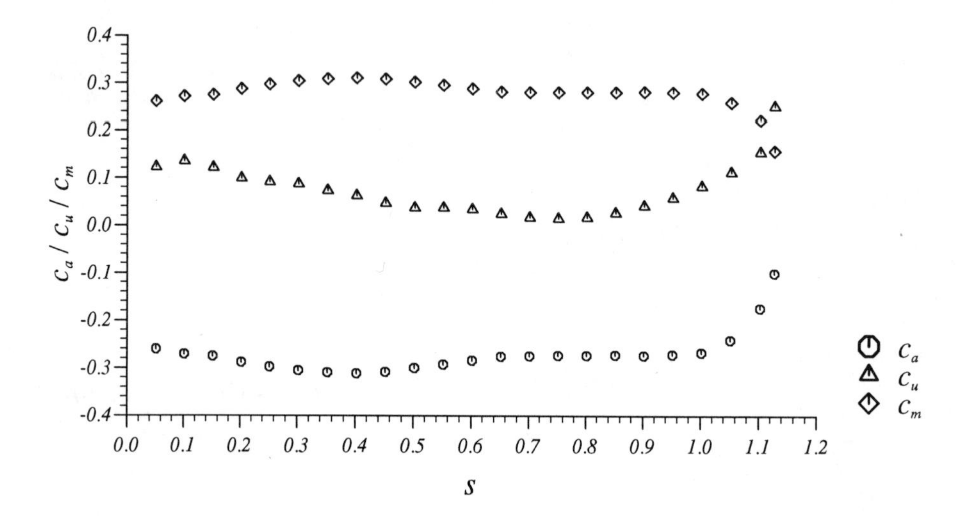

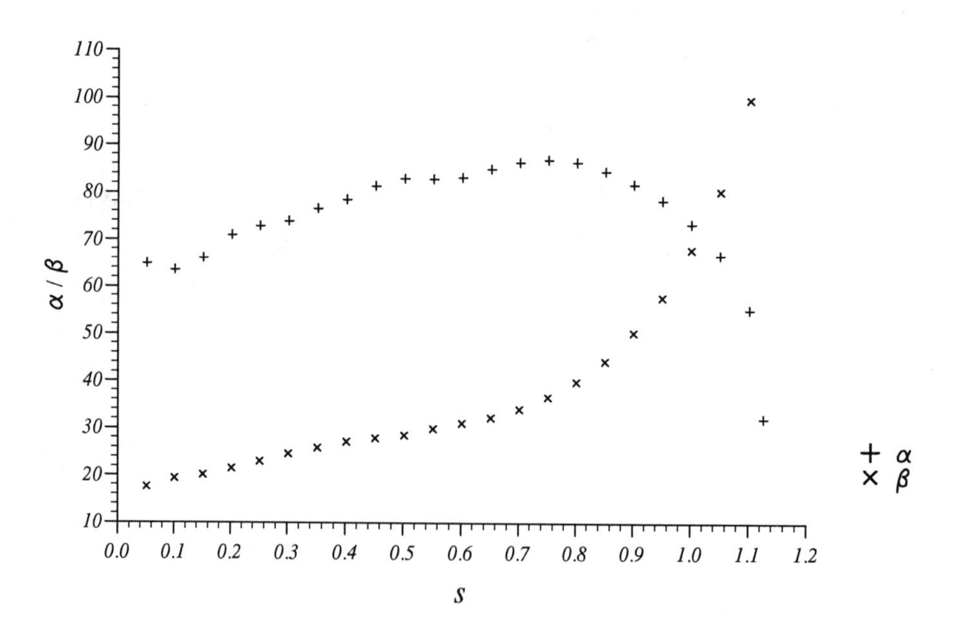

Fig. 12. Experimental velocity components and flow angles at the runner middle axis.

Requirement 1

c_p^* along the sections 2, 4, 6, 9, 12 and 15 of the blade versus the normalized arclength along the section's camber line.

$\{ N_{ps}, \{ N, \{R, \theta, Z, s^*, (c_p^*)_i, (c_p^*)_e\}_{N_{ps}} \}_N \}$

N_{ps}	number of camber line points per sections
N	number of the considered section (integer belonging to the set { *2, 4, 6, 9, 12, 15* }
$\{ \{R, \theta, Z\}_{N_{ps}} \}_N$	cylindrical coordinates of the camber line points per sections
$\{ \{s^*\}_{N_{ps}} \}_N$	normalized arclength corresponding to every camber line point
$\{ \{(c_p^*)_i, (c_p^*)_e\}_{N_{ps}} \}_N$	normalized pressure coefficient at the blade points corresponding to every camber line point (one on the pressure side and one on the suction side of the blade)

Note that : N_{ps} and $\{ \{R, \theta, Z\}_{N_{ps}} \}_N$ are part of the given geometrical data.

Let one of the above sections of the blade be chosen ($N = 2, 4, 6, 9, 12$ or 15) and let (R, θ, Z) be one point of the camber line of this section. The arclength s associated with the point (R, θ, Z) is the length of the portion of the camber line joining the leading edge point of this section to the point (R, θ, Z) and is computed assuming that the camber line is piecewise linear (constituted of $(N_{ps} - 1)$ segments). The associated normalized arclength s^* is then simply the arclength s divided by R_{ref}.

Two blade points are naturally associated with the camber line point (R, θ, Z), one on the pressure side and one on the suction side of the blade (they are constructed by using the normal vector and the width of the profile associated with the camber line point). $(c_p^*)_i$ and $(c_p^*)_e$ are the normalized pressure coefficients evaluated at these two points.

Requirement 2

averaged c_p^*, c_a^*, c_u^*, c_m^*, α and β along the runner inlet and outlet measurement axes versus the normalized arclength along the measurement axis.

$\{ \{ I, N_I, \{R, Z, s^*, (c_p^*)_{av}, (c_a^*)_{av}, (c_u^*)_{av}, (c_m^*)_{av}, \alpha, \beta\}_{N_I} \}_I \}$

I	indicator of the measurement axis (which value belongs to the set { *inlet, outlet* })
N_I	number of points per measurement axis
$\{ \{R, Z\}_{N_I} \}_I$	cylindrical coordinates of the measurement points

$\{\ \{s^*\}_{N_I}\ \}_I$	normalized arclength corresponding to every measurement point
$\{\ \{\ (c_p^*)_{av}\ \}_{N_I}\ \}_I$	averaged normalized pressure coefficient at each measurement point
$\{\{(c_a^*)_{av},(c_u^*)_{av},(c_m^*)_{av}\}_{N_I}\}_I$	averaged normalized velocity components at each measurement point
$\{\ \{\alpha,\ \beta\}_{N_I}\ \}_I$	absolute and relative flow angle corresponding to the averaged velocity components

Note that : N_I and $\{\ \{R, Z\}_{N_I}\ \}_I$ are part of the given measured data.

Let one of the measurement axis be chosen (I = inlet or outlet) and let (R, Z) be one of the given points on it. The arclength s associated with the point (R, Z) is the length of the portion of axis joining the band point of the axis to the point (R, Z). The associated normalized arclength s^* is then simply the arclength s divided by R_{ref}. Note here that the intersection of the inlet, respectively the outlet, axis with the band is at R = 210.9 [mm] and Z = –62.78 [mm], respectively R = 218.38 [mm] and Z = –346.35 [mm].

Since the measurement axes have a fixed position (i.e. they do not rotate with the runner), the measured data are averaged values, the averaging being on circles of radius R and located at height Z. Thus the computed c_p^*, c_a^*, c_u^*, c_m^* should be averaged in the same way in order to produce quantities which can be compared meaningfully with the measured data.

Requirement 3

averaged c_p^*, c_a^*, c_u^*, c_m^*, α and β along the runner middle measurement axis versus the normalized arclength along the measurement axis.

$$\{\ N_{pa},\ \{s^*,\ (c_p^*)_{av},\ (c_a^*)_{av},\ (c_u^*)_{av},\ (c_m^*)_{av},\ \alpha,\ \beta\}_{N_{pa}}\ \}$$

N_{pa}	number of points per measurement axis
$\{s^*\}_{N_{pa}}$	normalized arclength corresponding to every point on the measurement axis
$\{\ (c_p^*)_{av}\ \}_{N_{pa}}$	averaged normalized pressure coefficient at each point on the measurement axis
$\{(c_a^*)_{av},(c_u^*)_{av},(c_m^*)_{av}\}_{N_{pa}}$	averaged normalized velocity components at each point on the measurement axis
$\{\alpha,\ \beta\}_{N_{pa}}$	absolute and relative flow angle corresponding to the averaged velocity components

Note that : because we did not know the exact location of the measurement points on this axis before the workshop, we have fixed arbitrary points where the quantities have to be evaluated. These are points with normalized arclength equal to n∗0.05 where n is an integer between 1 and 21 (i.e. $N_{pa} = 21$).

The arclength s associated with a point (R, Z) on the axis is the length of the portion of axis joining the band point of the axis to the point (R, Z). The associated normalized arclength s^* is then simply the arclength s divided by R_{ref}. Note here that the intersection of the middle axis with the band is at $R = 205.09$ [mm] and $Z = -208.47$ [mm].

The reason of requiring averaged values for the pressure coefficient and for the velocity components is the same as for Requirement 2.

Requirement 4

c_p^*, c_a^*, c_u^*, c_m^*, α and β along the leading and the trailing edges of the blade versus the normalized arclength along the edge.

$$\{ N_s, \{I_e, \{R, \theta, Z, s^*, c_p^*, c_a^*, c_u^*, c_m^*, \alpha, \beta\}_{N_s} \}_{I_e} \}$$

N_s	number of sections defining the blade (i.e. number of points defining the leading or the trailing edge of the blade)
I_e	indicator of the edge of the blade (which value belongs to the set { *leading*, *trailing* })
$\{ \{R, \theta, Z\}_{N_s} \}_{I_e}$	cylindrical coordinates of the blade edge points
$\{ \{s^*\}_{N_s} \}_{I_e}$	normalized arclength corresponding to every blade edge point
$\{ \{ c_p^* \}_{N_s} \}_{I_e}$	normalized pressure coefficient at each blade edge point
$\{ \{ c_a^*, c_u^*, c_m^* \}_{N_s} \}_{I_e}$	normalized velocity components at each blade edge point
$\{ \{\alpha, \beta\}_{N_s} \}_{I_e}$	absolute and relative flow angle corresponding to the velocity components

Note that : N_s and $\{ \{R, \theta, Z\}_{N_s} \}_{I_e}$ are part of the given geometrical data.

Let one of the edges be chosen (I_e = leading or trailing), and let (R, θ, Z) be one of the given points on it. The arclength s associated with the point (R, θ, Z) is the length of the portion of the edge joining the point (R, θ, Z) to the edge point leading on the section N_s and is computed assuming the edge line is piecewise linear (constituted of $(N_s - 1)$ segments). The associated normalized arclength s^* is then simply the arclength s divided by R_{ref}.

Requirement 5

global physical quantities : t^* and η.

t^*	resultant nondimensional torque
η	efficiency

Note that : evaluate these two physical quantities by using the computed T (do not use the reference value given in the final announcement).

Requirement 6

computer related quantities :

- computer type,
- number of grid points used,
- cpu time,
- (cpu time) / iteration,
- (cpu time) / (number of grid cells),
- (cpu time) / (iteration * number of grid cells),
- ratio of vectorisation : cpu time (scalar version) / cpu time (vectorised version),
- parallelisation speedup,
- convergence criterion and convergence diagrams (if an iterative approach is used).

TEST CASE 3 : DRAFT TUBE

DRAFT TUBE GEOMETRY

The domain for the flow analysis in the draft tube will be in between the section containing the horizontal measurement axis towards the upstream side, and the outlet section of the draft tube towards the downstream side.

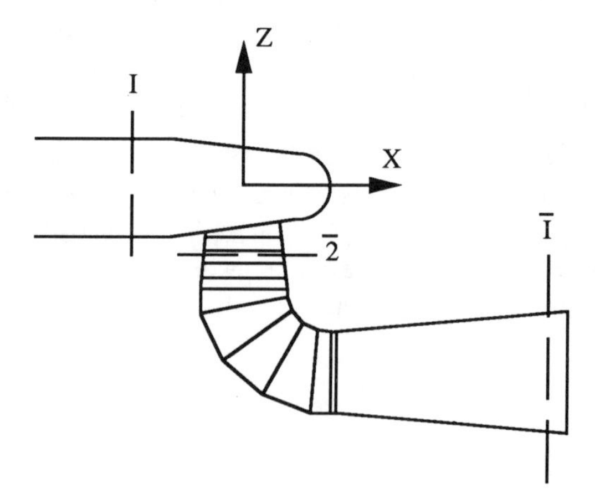

Fig. 13. Schematic view of the turbine showing the draft tube and different reference sections where the hydraulic energy is measured

The draft tube is an assembly of cone or cylinder portions (see Figure 13); thus it is specified using sections. The geometrical data are specified as follows :

Draft tube :	$\{ N_s, N_{ps}, \{ \{X, Y, Z\}_{N_{ps}} \}_{N_s} \}$	
	N_s	number of circular sections from measurement axis to outlet
	N_{ps}	number of points per circular sections
	$\{ \{X, Y, Z\}_{N_{ps}} \}_{N_s}$	cartesian coordinates of the points per sections

PHYSICAL DATA PROVIDED FOR THE FLOW SIMULATION IN THE DRAFT TUBE

The following data are given :

• the specific hydraulic energy between ref. sections I & $\overline{2}$	$E_{I \div \overline{2}}$	=	58.42	[J/kg]
• the specific hydraulic energy between ref. sections I & $\overline{I}$	$E_{I \div \overline{I}}$	=	60.33	[J/kg]
• the reference radius	R_{ref}	=	0.200	[m]
• the reference static pressure	P_{ref}	=	94300.000	[N/m²]
• the mean static pressure at the outlet of the draft tube	P_{dtout}	=	101566.000	[N/m²]
• the gravity constant	g	=	9.806	[m/s²]
• the flow rate	$\dot{V}$	=	0.372	[m³/s]
• the water density	ρ	=	1000.000	[kg/m³]

- the pressure distribution $P - P_{ref}$ measured at the inlet of the draft tube
- the velocity distribution (C_R, C_θ, C_Z) measured at the inlet of the draft tube

The pressure distribution $P - P_{ref}$ and the velocity distribution (C_R, C_θ, C_Z) measured at the inlet of the draft tube, together with the location of the measurement points, are specified in the following way :

Inlet data :	$\{ N_a, N_{pa}, \{ \{R, \theta, Z, P - P_{ref}, C_R, C_\theta, C_Z\}_{N_{pa}} \}_{N_a} \}$	
	N_a	number of measurement axis
	N_{pa}	number of points per measurement axis
	$\{ \{R, \theta, Z\}_{N_{pa}} \}_{N_a}$	cylindrical coordinates of the measurement points
	$\{ \{P \; P_{ref}\}_{N_{pa}} \}_{N_a}$	pressure at each measurement points
	$\{ \{C_R, C_\theta, C_Z\}_{N_{pa}} \}_{N_a}$	velocity components at each measurement points

These data, in their normalized form are presented in Figure 14.

Since the inlet axis does not rotate with the runner, the measured data are averaged values.

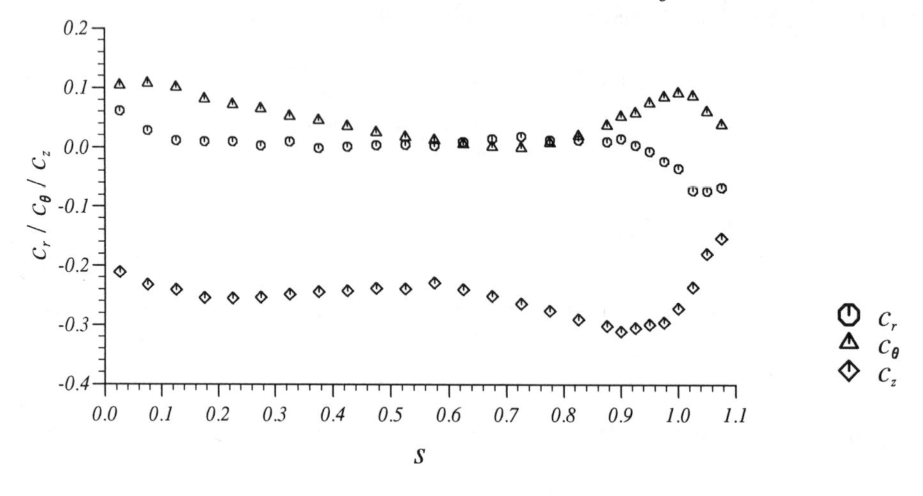

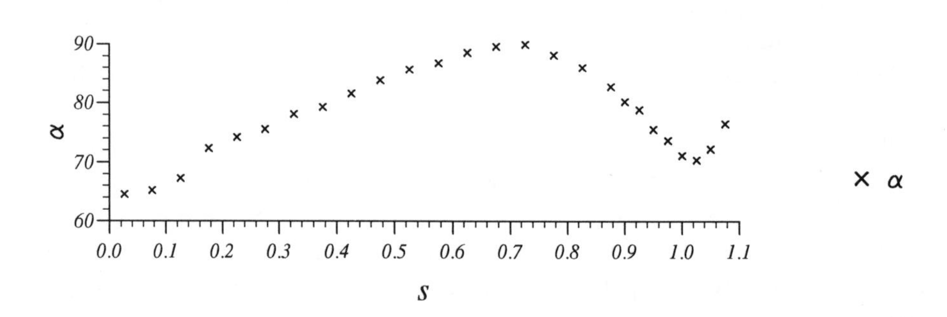

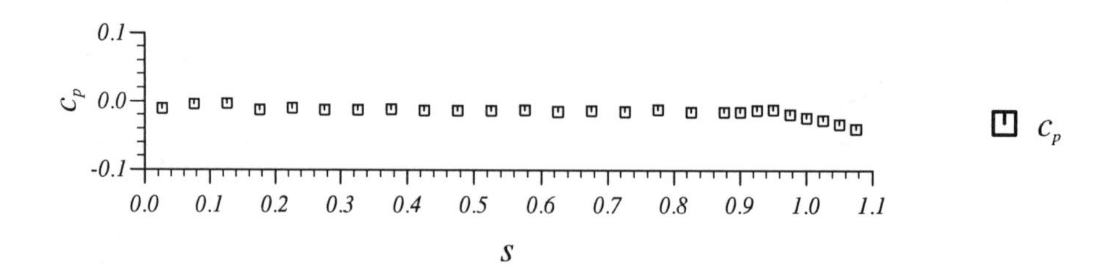

Fig. 14. Experimental velocity components, flow angle and pressure coefficient at the draft tube inlet axis.

Required results for the draft tube flow

We recall here that the geometry of the draft tube has been defined using N_s (= 12) circular or elliptic sections. This fact is used in the description of the required results given below.

Requirement 1

c_p^* along the draft tube axis versus the normalized arclength along the axis.

$\{ N_s, \{X, Y, Z, s^*, c_p^*\}_{N_s} \}$

N_s	number of circular or elliptic sections from measurement axis to outlet
$\{X, Y, Z\}_{N_s}$	cartesian coordinates of the centers of the sections (i.e. coordinates of the points defining the draft tube axis)
$\{s^*\}_{N_s}$	normalized arclength corresponding to every axis point
$\{c_p^*\}_{N_s}$	normalized pressure coefficient at each axis points

Note that : N_s is part of the given geometrical data.

The draft tube axis is a piecewise linear curve constituted of $(N_s - 1)$ segments. The end points of each segment are the centers of the given sections.

Let (X, Y, Z) be one point of the draft tube axis. The arclength s associated with the point (X, Y, Z) is the length of the portion of the draft tube axis joining the axis point belonging to section 1 to the point (X, Y, Z). The associated normalized arclength s^* is then simply the arclength s divided by $12*R_{ref}$ (= 2400 [mm]).

Requirement 2

averaged c_p^*, c_R^*, c_θ^*, c_Z^*, and α along the draft tube inlet measurement axis versus the normalized arclength along the measurement axis.

$\{ N_{pa}, \{R, Z, s^*, (c_p^*)_{av}, (c_R^*)_{av}, (c_\theta^*)_{av}, (c_Z^*)_{av}, \alpha\}_{N_{pa}} \}$

N_{pa}	number of points per measurement axis
$\{R, Z\}_{N_{pa}}$	cylindrical coordinates of the measurement points
$\{s^*\}_{N_{pa}}$	normalized arclength corresponding to every measurement point
$\{ (c_p^*)_{av} \}_{N_{pa}}$	averaged normalized pressure coefficient at each measurement point

$\{(c_R^*)_{av},(c_\theta^*)_{av},(c_Z^*)_{av}\}_{N_{pa}}$	averaged normalized velocity components at each measurement point
$\{\alpha\}_{N_{pa}}$	absolute flow angle corresponding to the averaged velocity components

Note that : N_{pa} and $\{R, Z\}_{N_{pa}}$ are part of the given measured data.

Let (R, Z) be one of the given points on the measurement axis. The arclength s associated with the point (R, Z) is the length of the portion of axis joining the point (R, Z) to the axis point located at $R = 218.38$ [mm] and $Z = -346.35$ [mm]. The associated normalized arclength s^* is then simply the arclength s divided by R_{ref}.

At the inlet section, the hypothesis of axisymmetry is made. Thus the computed values c_p^*, c_R^*, c_θ^*, c_Z^* should be averaged on circles of radius R and located at constant height Z.

Requirement 3

elevation view of the velocity vector field.

Consider the section of the draft tube determined by its intersection with the vertical (X,Z) plane. Note that, because of the definition of the coordinate system, the draft tube axis belongs to this section. Display the velocity field on this section.

Requirement 4

velocity vector field on the cross sections 1, 8, 11 and 12.

Note that : these cross sections are part of the sections defining the geometry.

Requirement 5

computer related quantities :

- computer type,
- number of grid points used,
- cpu time,
- (cpu time) / iteration,
- (cpu time) / (number of grid cells),
- (cpu time) / (iteration * number of grid cells),
- ratio of vectorisation : cpu time (scalar version) / cpu time (vectorised version),
- parallelisation speedup,
- convergence criterion and convergence diagrams (if an iterative approach is used).

Part 2

EXPERIMENTAL RESULTS

EXPERIMENTAL FLOW STUDY OF THE GAMM TURBINE MODEL

François Avellan, Philippe Dupont, Mohamed Farhat, Bernard Gindroz, Pierre Henry, Mahmood Hussain, Etienne Parkinson, Olivier Santal

EPFL, Institut de Machines Hydrauliques et de Mécanique des Fluides
33, avenue de Cour, CH 1007 Lausanne, Switzerland

SUMMARY

An experimental study of flow in a Francis hydraulic turbine, especially designed for the GAMM Workshop, has been carried out. In order to provide a data base for the validation and comparison of computational fluid dynamics codes, a Francis turbine model and adequate instrumentation have been specially developed. Measurements of both global and local quantities have been performed for the best operating point of the turbine. A five-hole probe mounted on a remote traversing system provides static pressure and velocity components along three measurement axes in the machine. Pressure transducers mounted on the pressure and suction sides of the rotating blades of the runner provide the pressure distribution over the blades in 28 points (17 on the suction side and 11 on the pressure side) arranged along three theoretical streamlines.

INTRODUCTION

For the GAMM workshop on the computation of incompressible internal flows, IMHEF has provided a full geometry of a Francis turbine. A model Francis turbine was built according to the specified geometry and tested in the IMHEF test facilities, in order to compare the results produced by various CFD codes used by the participants with the measured data. These tests provided as well a proper specification of the test case flow conditions. In this experimental model setup, special features were included, such as, an adjustable stayring, probe passages and pressure taps on the stationary parts. The runner blades were equipped with pressure transducers for measuring the pressure distribution on the blade surfaces.

The aim of this paper is to provide a brief description of the experimental setup used for studying the flow pattern in a model Francis turbine. Special attention was paid to the design of the flow survey instrumentation consisting of a five-hole probe mounted on a remote traversing system, and the pressure measurement system used in the rotating runner. Detailed results of the flow survey in three different locations of the model are given for the best operating point during the tests carried out using the IMHEF Universal Hydraulic Machine Test Facility [1].

THE MODEL OF FRANCIS TURBINE

GENERAL DESIGN

The test model corresponds to a Francis turbine of medium-high specific speed, $\nu = 0.5$, ($n_q = 76$). This model was specially designed by IMHEF only for research purposes in order to perform experimental flow studies. A Piguet-type spiral case was designed to give a constant meridional velocity distribution. A fillet was added on both sides of the stayring inlet in order to assure well-defined inflow conditions.

The stayring consists of 24 stay-vanes and the distributor of 24 guide-vanes. The relative angular position of these two cascades is adjustable, but, for the present study, the angular positions θ_d and θ_{ad} remained unchanged and were set to 26.5° and 6.5° respectively, see Figure 1. A mechanical encoder provides a readout of the guide vane opening angle α. Since the closed position of the guide vanes corresponds to $\alpha = 0°$ and an angle $\gamma_{d0} = 5.37°$, as indicated in Figure 1, the opening angle α is related to the guide vane angle as follows :

$$\gamma_d = \alpha + \gamma_{d0}.$$

The runner has 13 blades, each individually casted in epoxy resin reinforced with carbon fiber. The blades are fixed in between an aluminium hub and shroud. The runner outlet diameter is 0.4 m. Top view of the horizontal cross-sections of both the pressure and the suction sides of the blade are given in Figure 2.

The draft tube has a simple shape. It was designed to provide for numerical simulation work a simplified geometry without any inner pillar. It consists of an inlet cone, a constant section bend and an outlet cone. The geometry of the draft tube is shown in Figure 3.

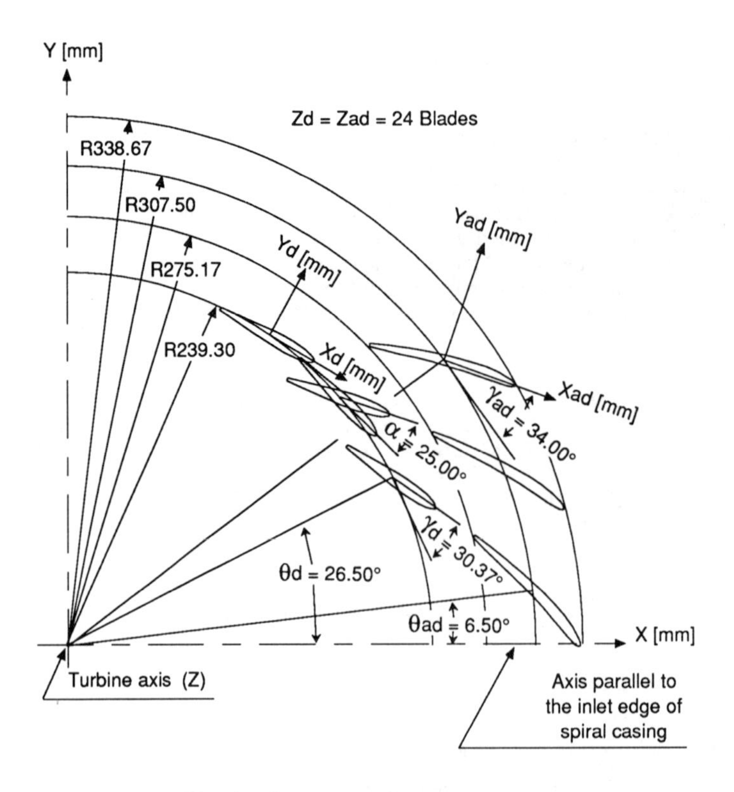

Fig. 1 Stayring and guide-vanes, top view

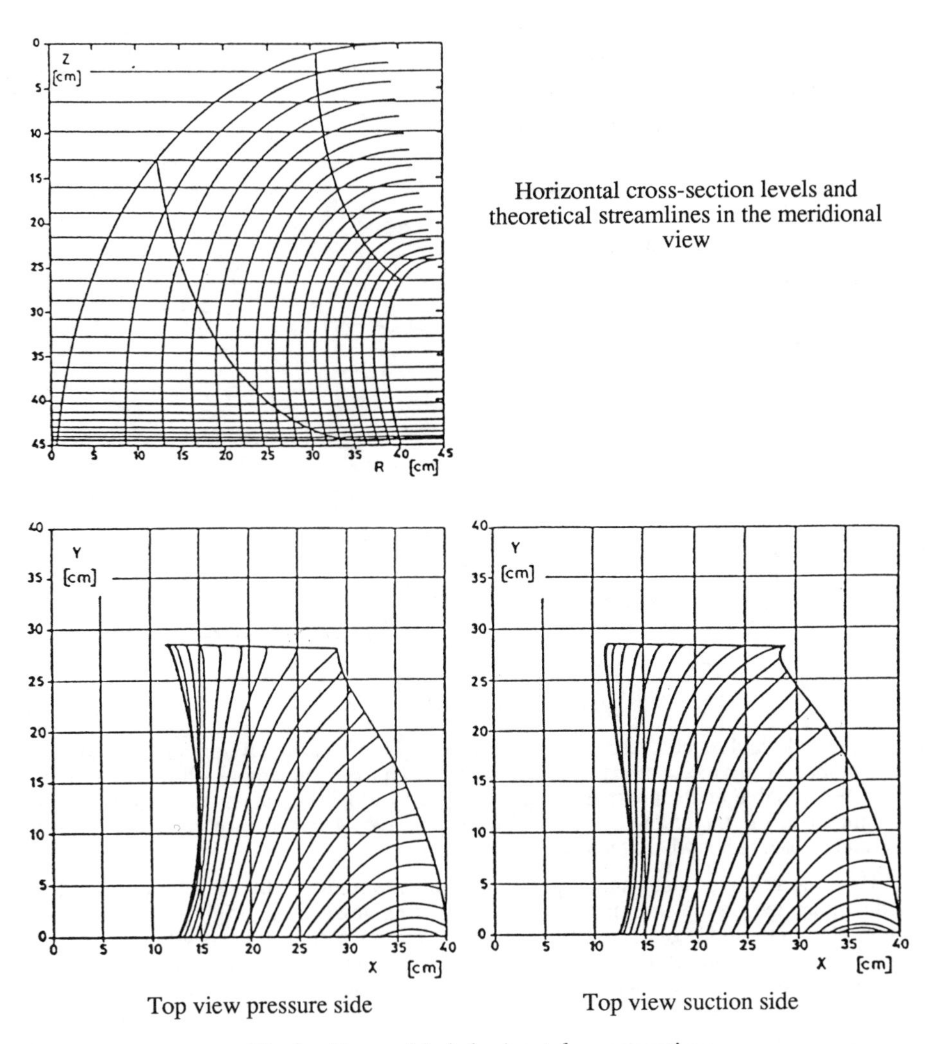

Horizontal cross-section levels and theoretical streamlines in the meridional view

Top view pressure side

Top view suction side

Fig. 2 Runner blade horizontal cross-sections

FLOW SURVEY SYSTEM

More than 60 probe passages in the model were planned in order to perform a complete flow survey. They are placed at the stayring inlet, at the runner inlet and outlet, and in a cross-section of the plexi-glass cone. The location of these passages and their corresponding measurement axis are defined in Figure 4. A special mechanical arrangement allows to convert these probe passages into emplacements for static pressure taps.

A remote traversing system capable of transmitting a rotatory or a translatory movement to the probe is mounted on the model base-plate in order to have an access to any desired measuring point in the fixed flow passages of the model. The traversing system support is made up of a rigid light alloy square tube, see Figure 5. The turntable, which allows the angular positioning, is mounted in one end of the tube. The turntable is supported

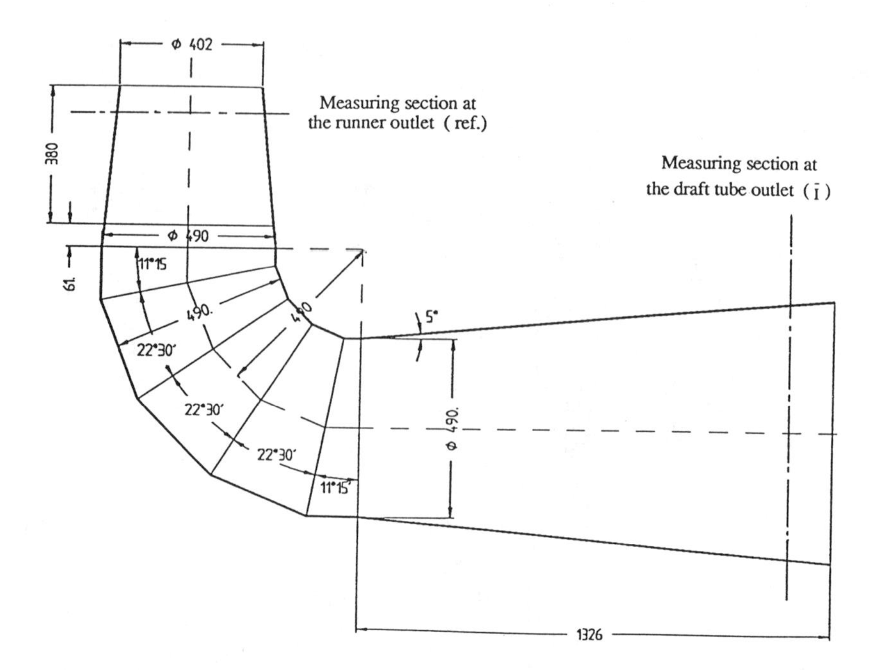

Fig. 3 Draft tube geometry

by a preloaded playless ball race and driven by a step motor with a worm and gear system. Linear motion is produced by using a step motor driven actuator mounted on the rotating part of the turntable and supported at the other end by a ball-bearing fitted into the square tube support of the traversing system. The linear drive system incorporates a fixed lead screw and a moving nut where the probe support is mounted.

Both the relative linear and angular positions are numerically read by optical shaft encoders mounted on the step motor shafts. To define an accurate origin for each axis of motion, electronic logic gates are used, firstly to rectify a contactless switch output and secondly to combine this output with the corresponding encoder signal output. The resulting signal is used to permit the resetting of the electronic counters. The sensitivity of one motor step corresponds to 6×10^{-6} m and 10^{-2} ° for the translational and the rotational motions, respectively. An accuracy of less than $\pm 20 \times 10^{-6}$ m and $\pm 0.2°$ is achieved in the probe positioning. The linear positioning capacity is 0.3 m and the probe can be rotated through 360°.

THE INSTRUMENTATION OF THE RUNNER

The pressure transducers were mounted flush with the blade surfaces of the runner. The pressure transducers (resistive bridge) have been fitted on the model of the blades during the casting process. The wires of each full bridge were embedded in the carbon fiber and epoxy resin of the blades up to the hub of the runner. All the elements of processing and transmitting electronics were placed in the hub of the runner. The lead and signal wires were passed through the hollow shaft up to the transmitting coil. This assembling process allowed to place 3 transducers on each blade. In total 28 pressure transducers were installed on different blades, 17 of them were mounted on the suction side of the blades, and 11 on the pressure side, see Figure 6.

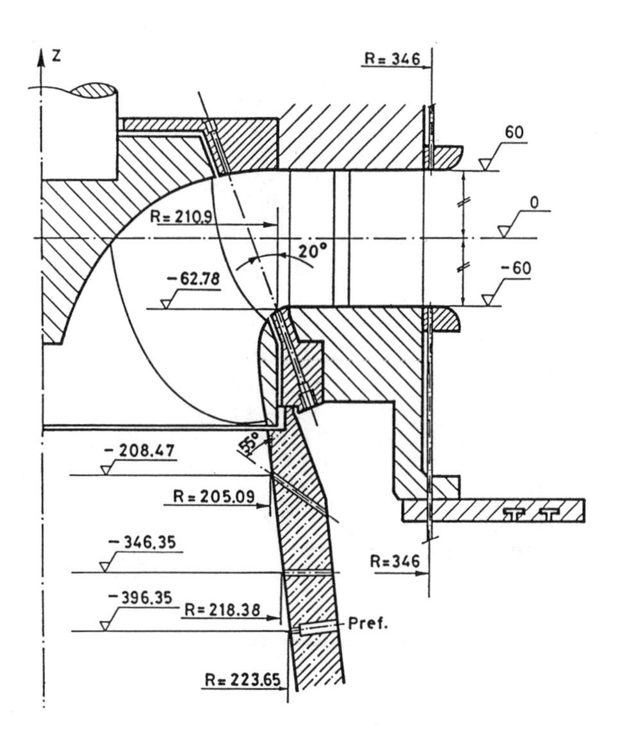

Fig. 4 Flow survey axes

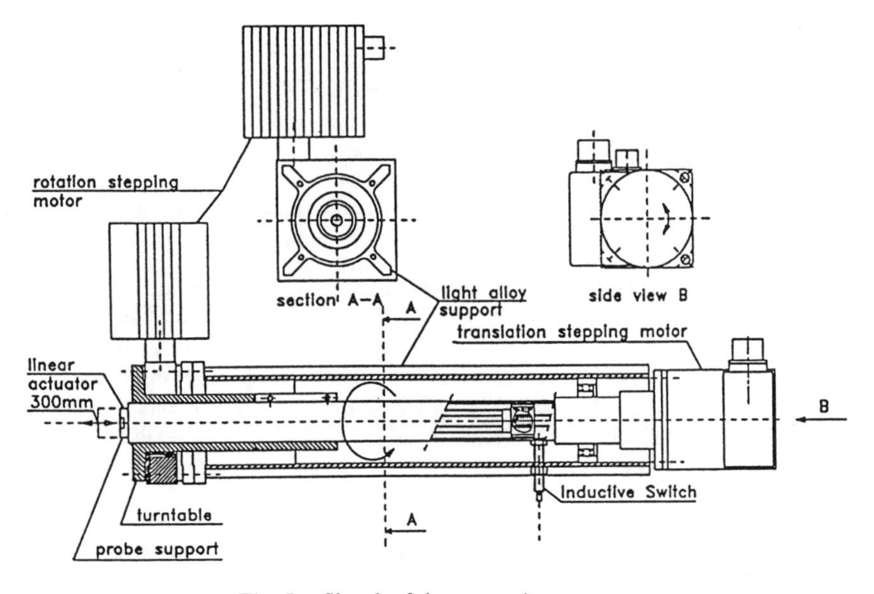

Fig. 5 Sketch of the traversing system

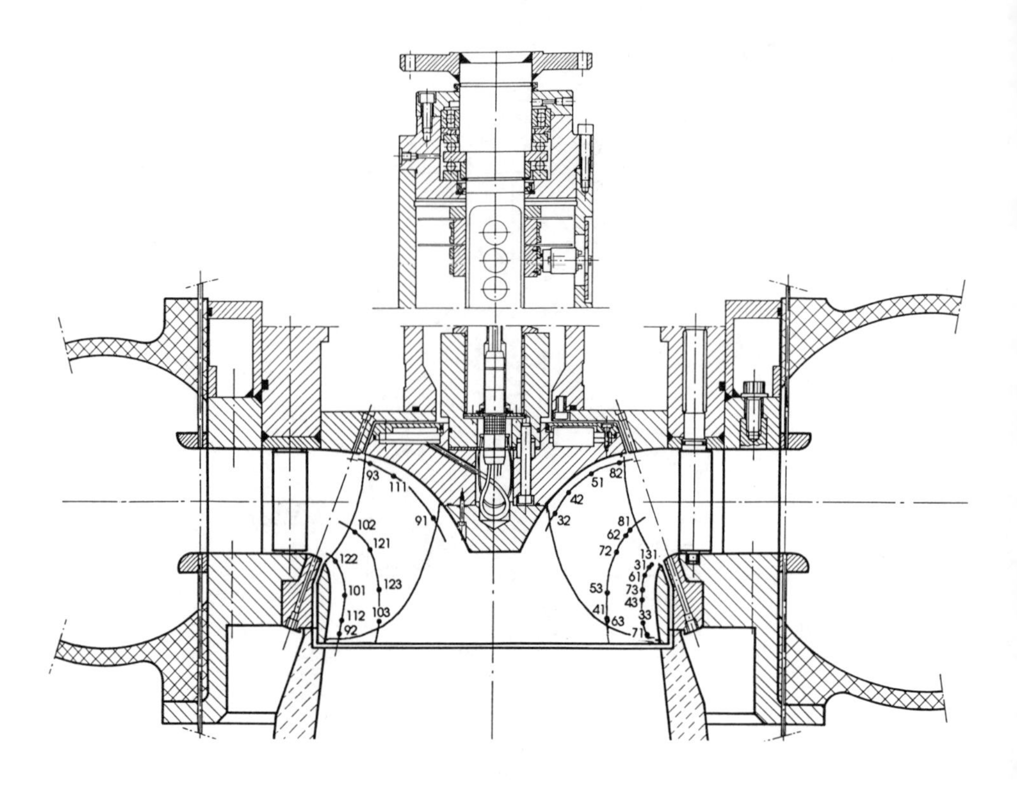

Fig. 6 Instrumented runner

PERFORMANCE TESTS

HYDRAULIC CHARACTERISTICS

Model tests were performed in the IMHEF Universal Hydraulic Machine Test Facility. Its principal characteristics are : a variable speed main pump of 900 kW at 1'000 rpm, capable of furnishing a maximum flow rate of 1.4 m^3/s, a test head range of 2-100 m with a maximum dynamometer power of 300 kW at a maximum speed of 2'500 rpm. The accuracy of the test instrumentations is far better than the IEC model turbine acceptance test code requirements.

The hydraulic characteristics for a given guide-vane opening angle are given in a normalized φ-ψ form, see Figure 7. The model efficiency hillchart reported in the same figure presents two maxima caused by the poor pressure recovery of the conical draft tube.

CONE REFERENCE

The specific hydraulic energy was measured using two different downstream sections as shown in Figure **3** :

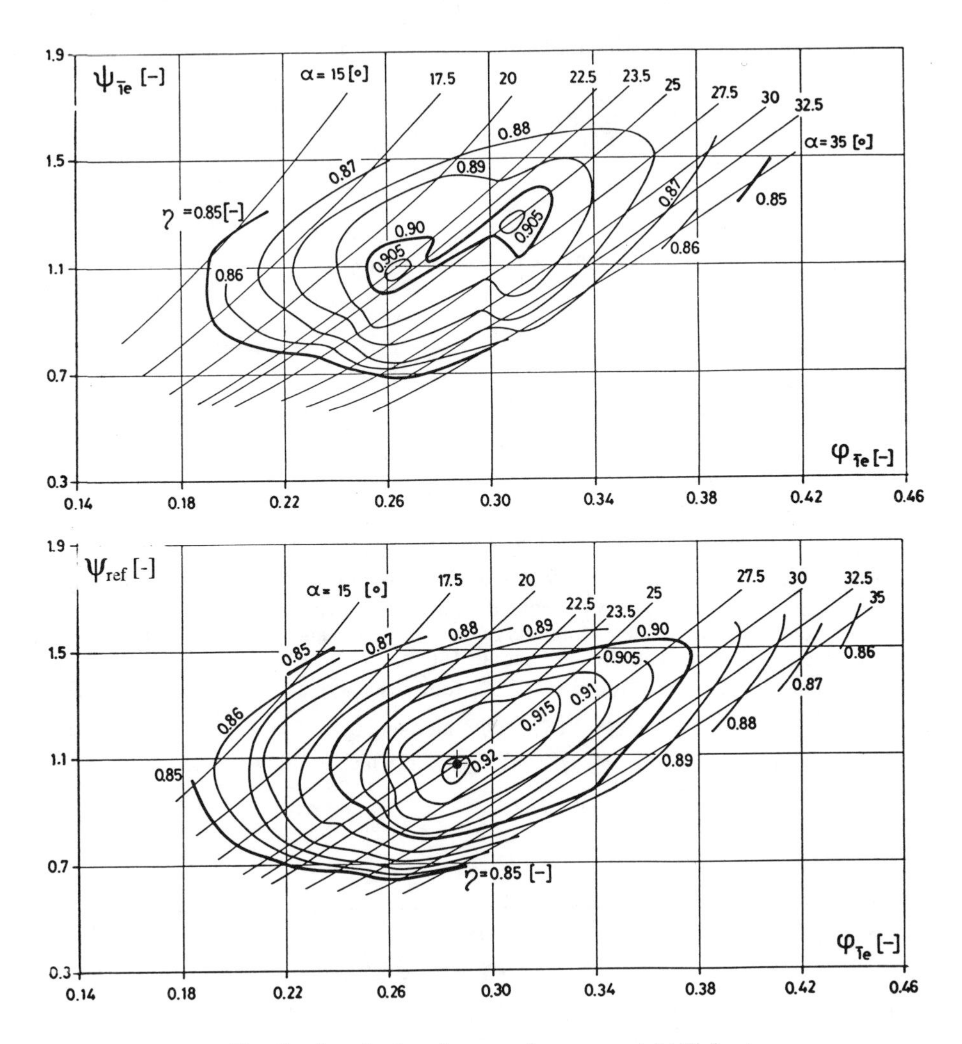

Fig. 7 Standard and cone reference model hillcharts

- the first (IEC defined) standard measuring section ($\bar{I}$) is placed at the outlet of the draft-tube,
- the second (IMHEF defined) measuring section (ref) is close to the runner outlet.

The standard (IEC) specific hydraulic energy is given by

$$E = \frac{p_I}{\rho} + \frac{Q_I^2}{2\,S_I^2} + gZ_I - \frac{p_{\bar{I}}}{\rho} - \frac{Q_{\bar{I}}^2}{2\,S_{\bar{I}}^2} - gZ_{\bar{I}}$$

where the standard section (I) is taken at the machine inlet.

The specific hydraulic energy using the runner outlet measuring section is given by

$$E_{ref} = \frac{p_I}{\rho} + \frac{Q_I^2}{2\,S_I^2} + gZ_I - \frac{p_{ref}}{\rho} - \frac{Q_{ref}^2}{2\,S_{ref}^2} - gZ_{ref}\,.$$

The corresponding energy coefficients are defined as follows

$$\psi_{\bar{1}e} = \frac{2E}{\omega^2\,(R_{\bar{1}e})^2}\,,$$

$$\psi_{ref} = \frac{2E_{ref}}{\omega^2\,(R_{\bar{1}e})^2}\,.$$

With these reference values, the problem of the poor draft-tube behaviour which gave an unusual hillchart with two peaks is solved. Thus, the modified energy coefficient ψ_{ref} and the efficiency were computed using this reference station, and are reported as a function of the discharge coefficient in the hillchart in Figure 7. In this case the best efficiency operating point leads to an efficiency $\eta = 0.92$ for a discharge coefficient of $\varphi_{\bar{1}e_\wedge} = 0.286$, and an energy coefficient of $\psi_{ref} = 1.072$, with an opening angle of guidevanes $\alpha = 25°$, thereby, giving the value of $\gamma_d = 30.37°$, see Figure 1, which is slightly different from what was enounced, $\gamma_d = 28.25°$.

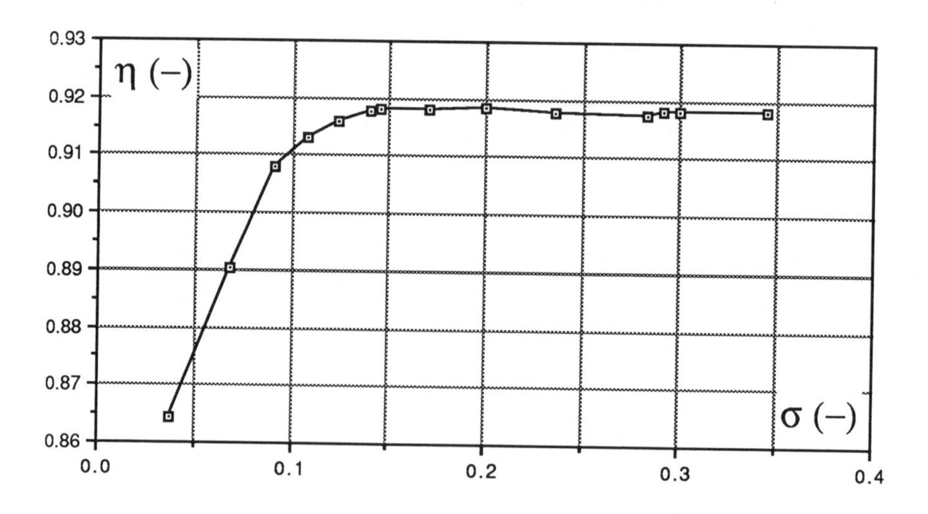

Fig. 8 Efficiency versus σ for the best efficiency point

Fig. 9 Photography of the inlet edge cavitation development at the best efficiency operating point

Cavitation

The cavitation behaviour of the runner is also very interesting since it provides a rough idea of the pressure distribution in the runner. Figure 8 gives the efficiency versus σ for the best efficiency operating point ($\varphi \text{I} e_{\wedge} = 0.286$, $\psi_{\text{re f}} = 1.072$).

The inlet edge cavitation at the suction side begins at $\sigma = 0.30$. Photography represented on Figure 9 shows the extension of inlet edge cavitation for $\sigma = 0.14$ and $\sigma = 0.20$. This cavity development is important and shows that the pressure near the inlet edge, close to the band on the suction side is equal to the vapour pressure below $\sigma = 0.30$.

FLOW SURVEY INSTRUMENTATION

Five-hole probe

Design

The 6 mm diameter five-hole probe was designed to perform a full traversing survey of the flow in the different stations of interest by giving the 3 components of the local flow velocity and the local static pressure. The arrangement of the 0.5 mm diameter pressure tap holes is the same as that of the United Sensor's probe, see Figure 10. In this case the probe is extended by a rod in order to guide it in the probe passages. The rod diameter is the same as that of the probe outer tube. The guide bearings in the probe passages eliminate mechanical vibrations as far as possible and allow the flow profile to be measured right across the flow channel for relevant head and flow values.

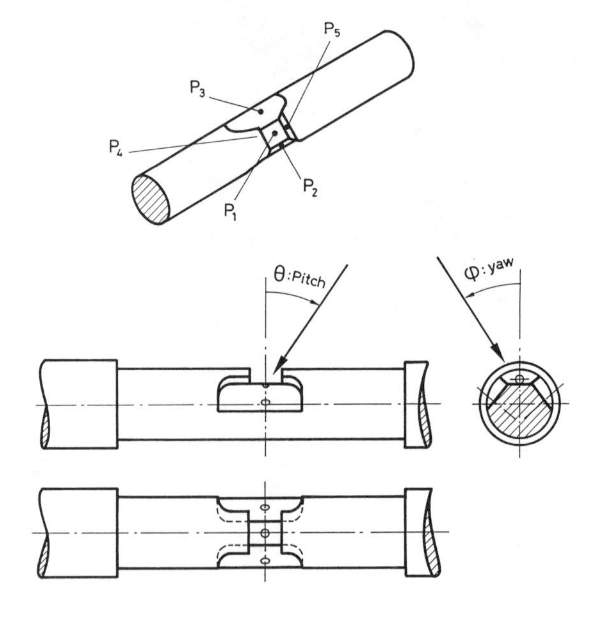

Fig. 10 Probe pressure tap arrangement

Static pressure measurements

Static pressures are measured by a water pressure line scanning system. This system, driven by a micro-computer, uses two Scanivalves connected to an absolute pressure transducer. The transducer is isolated from the facility water by silicone oil and automatic draining is performed as long as necessary to remove all the unwanted bubbles maintained in every pressure line. In addition to the five pressure taps of the probe, the pressure taps of the reference model inlet station, the cone station and all the other sets of pressure taps in the model are connected to the scanning line system.

Calibration

The probe was calibrated in the test section of the IMHEF High-Speed Cavitation Tunnel. Absolute pressure measurements corresponding to the five holes were carried out for different pitch and yaw angles of the probe as a function of the flow velocity, the total pressure, p_t, and the static pressure, p_o, in the test section. With the pressures p_2, p_3, p_4 and p_5, see Figure 10, one can define the average pressure p_m as follows :

$$p_m = (p_2 + p_3 + p_4 + p_5)/4 .$$

With this definition, the following coefficients F and G were introduced as functions of the yaw angle Φ and the pitch angle θ :

$$F(\Phi,\theta) = \frac{p_2 - p_3}{p_1 - p_m} \quad \text{and} \quad G(\Phi,\theta) = \frac{p_4 - p_5}{p_1 - p_m} ,$$

where the difference $p_1 - p_m$ behaves like a dynamic pressure in the range of the yaw angles.

Calibration curves are given in Figure 11 as contoured surfaces of F and G values in a Φ versus θ graph. It can be observed that F is almost independent of θ and the G coefficient is independent of Φ. In contrast to the F coefficient the G coefficient is not well-defined for the value of the yaw angle greater than 15° or smaller than -15°.

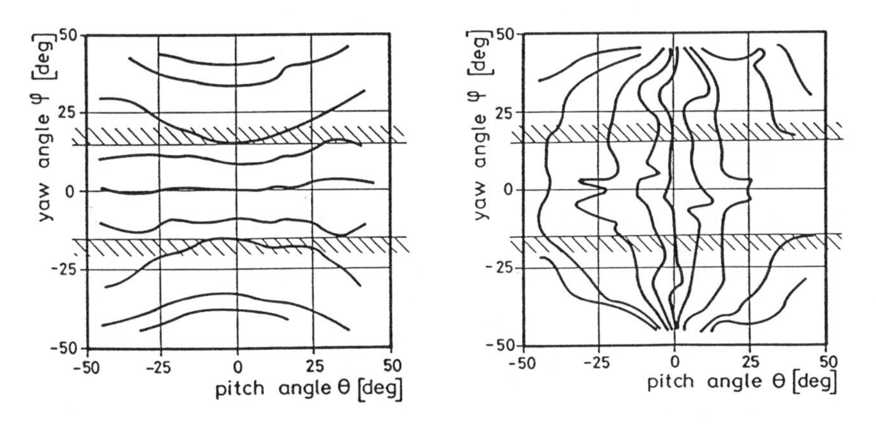

Fig. 11 Calibration surfaces of the yaw and pitch angles

In addition, the coefficients H and L can be defined as functions of the static pressure p_o and the total pressure p_t. The corresponding calibration curves are given in Figure 12 as contoured surfaces of H and L values in a Φ versus θ graph.

$$H(\Phi,\theta) = \frac{p_1 - p_m}{p_t - p_0} \quad \text{and} \quad L(\Phi,\theta) = \frac{p_1 - p_t}{p_t - p_0} .$$

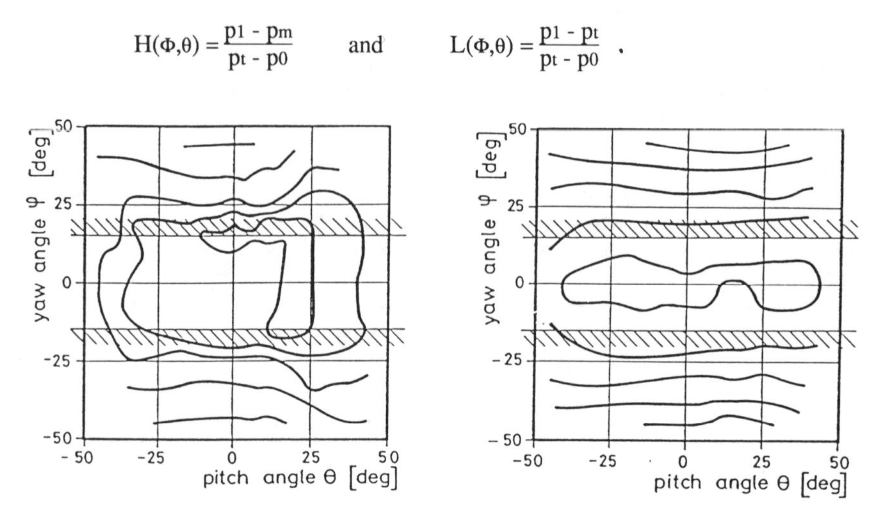

Fig. 12 Calibration surfaces of the total and static pressure

It is to be noted that, these strongly contoured curves represented in Figures. 11 and 12 are repetetive and the interpolated values are to be really read on the curves by successive approximation.

During the flow survey, the corresponding pressure values allow us to compute the F and G coefficients in order to find a corresponding pair of Φ, θ angles. The H and L coefficients are then calculated with the help of angle values, which, in turn lead to the static and the total pressure. The local velocity C is then computed according to the definition of the dynamic pressure :

$$C = \sqrt{\frac{p_t - p_0}{\frac{1}{2}\rho}} .$$

FLOW MEASUREMENT

The calibrated probe is mounted at the end of the linear actuator of the traversing system. The displacement origins are set using the switch references. The five probe pressures and those at the inlet reference and at the cone reference stations are measured sequentially. The flow angles are determined, and if the the yaw angle Φ is not within a range of $\pm 15°$, the probe is rotated into the specified range to achieve good accuracy for the flow velocity. The flow angle, the velocity intensity and static pressure are computed in the model frame of reference according to the definition in Figure 13.

Even though the actual flow rate value cannot be sampled simultaneously with the probe pressure measurement, this pressure scaling, related to the reference pressure using a value similar to a kinetic energy term, allows a comparison of the measurements, taken for different test heads. These coefficients are then related to the reference specific energy provided by the test-rig measuring system.

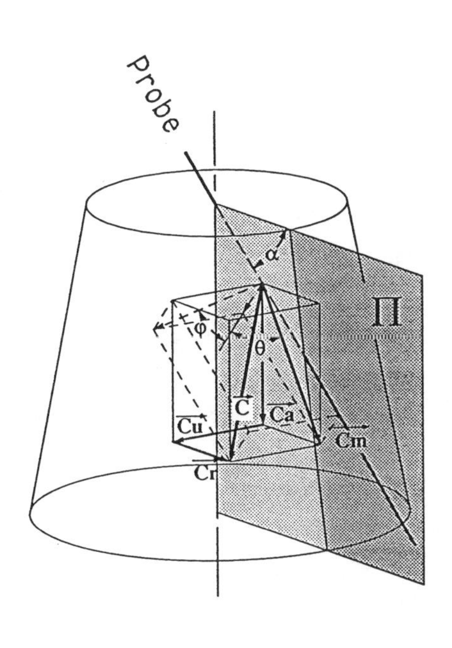

$$\begin{bmatrix} C_r \\ C_\theta \\ C_z \end{bmatrix} = \begin{bmatrix} C_r \\ C_u \\ -C_a \end{bmatrix}$$

$$= C \cdot \begin{bmatrix} -\sin\alpha & 0 & \cos\alpha \\ 0 & 1 & 0 \\ -\cos\alpha & 0 & -\sin\alpha \end{bmatrix} \cdot \begin{bmatrix} \cos\theta\cos\Phi \\ \cos\theta\sin\Phi \\ \sin\theta \end{bmatrix}$$

Fig. 13 Velocity component in the probe frame of reference. In this case α is the angle of the probe axis with the radial direction in the meridional plan

To overcome any drift conditions in the test facility operations during the flow survey, the actual static and total pressures are scaled by using a kinetic energy term defined as follows :

$$E_k = \frac{p_I}{\rho} + gZ_I - \frac{p_{ref}}{\rho} - gZ_{ref}$$

in order to compute the static pressure and total pressure coefficients, $C_{po(probe)}$ and $C_{pt(probe)}$ are defined as :

$$C_{po(probe)} = \frac{p_0 - p_{ref}}{\rho E_k} \quad \text{and} \quad C_{pt(probe)} = \frac{p_t - p_{ref}}{\rho E_k} .$$

Then the actual coefficient values are deduced by using the corresponding pair of kinetic energy term E_k and the specific energy E provided by the test rig panel control. Hence the coefficients are

$$C_{po} = C_{po(probe)} \frac{E_k}{E} \quad \text{and} \quad C_{pt} = C_{pt(probe)} \frac{E_k}{E} .$$

This leads to the following definition of the velocity coefficient c :

$$c = \frac{C}{\sqrt{2E}} .$$

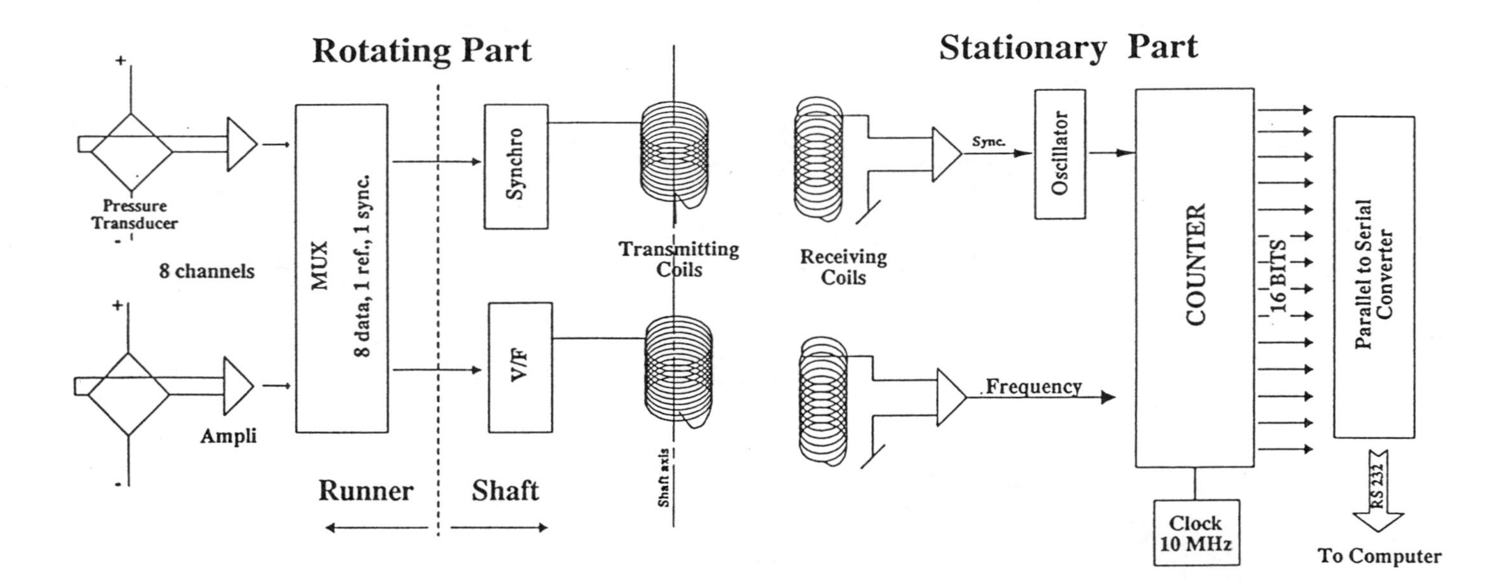

Fig. 14 *Conditioning and wireless transmitting electronics*

PRESSURE MEASUREMENTS IN THE RUNNER

The miniature absolute pressure transducers embedded in the runner blades are a Kyowa's resistive bridge-type PS-2KA model. Their diameter and thickness are 5 mm and 0.6 mm, respectively. The pressure range is 2×10^5 Pa and the mean sensitivity 900 µV/V. It should be noticed that the pressure should be always higher than the atmospheric pressure in order to avoid the destruction of transducers.

The multiplexed bridge signal is voltage to frequency converted in order to be transmitted through a coil mounted on the rotating shaft. The stationary part of the electronics consists of a receiving coil and a numerical counter driven through a serial interface by the microcomputer, Figure 14.

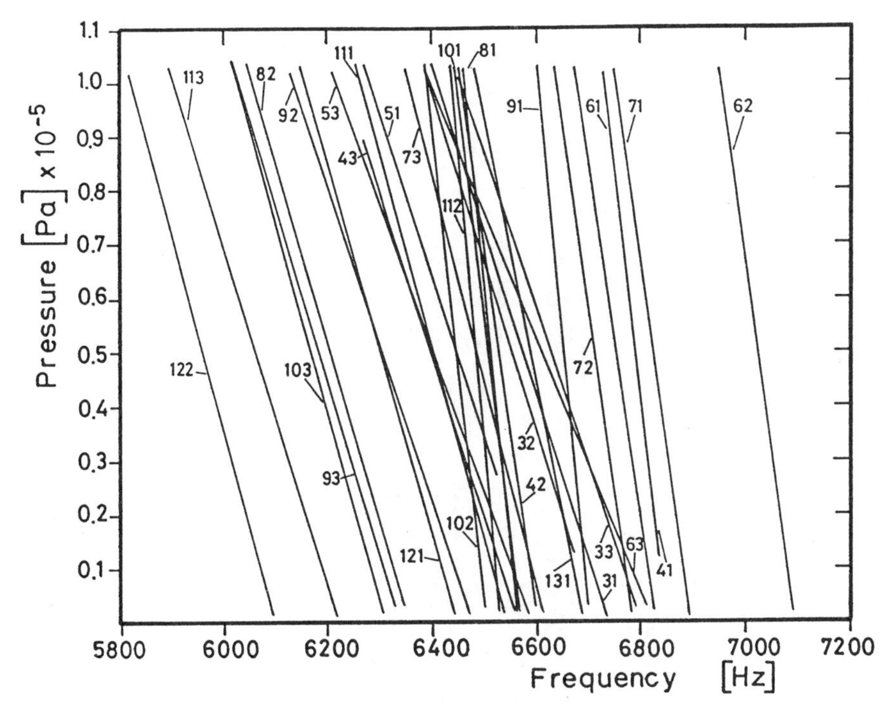

Fig. 15 Calibration curves of the rotating pressure transducers (see Figure. 6 for transducer identification)

When the runner is at rest, the pressure transducer is calibrated by varying the static pressure of the test-rig, Figure 15. The pressure distribution is obtained for a given operating point by scanning the 28 pressure transducers. The data is reduced by computing the time average of the pressure signal corresponding to the nth transducer and scaled as follows :

$$C_p^* = \frac{p_n - p_{ref}}{p_I - p_{ref}} \;.$$

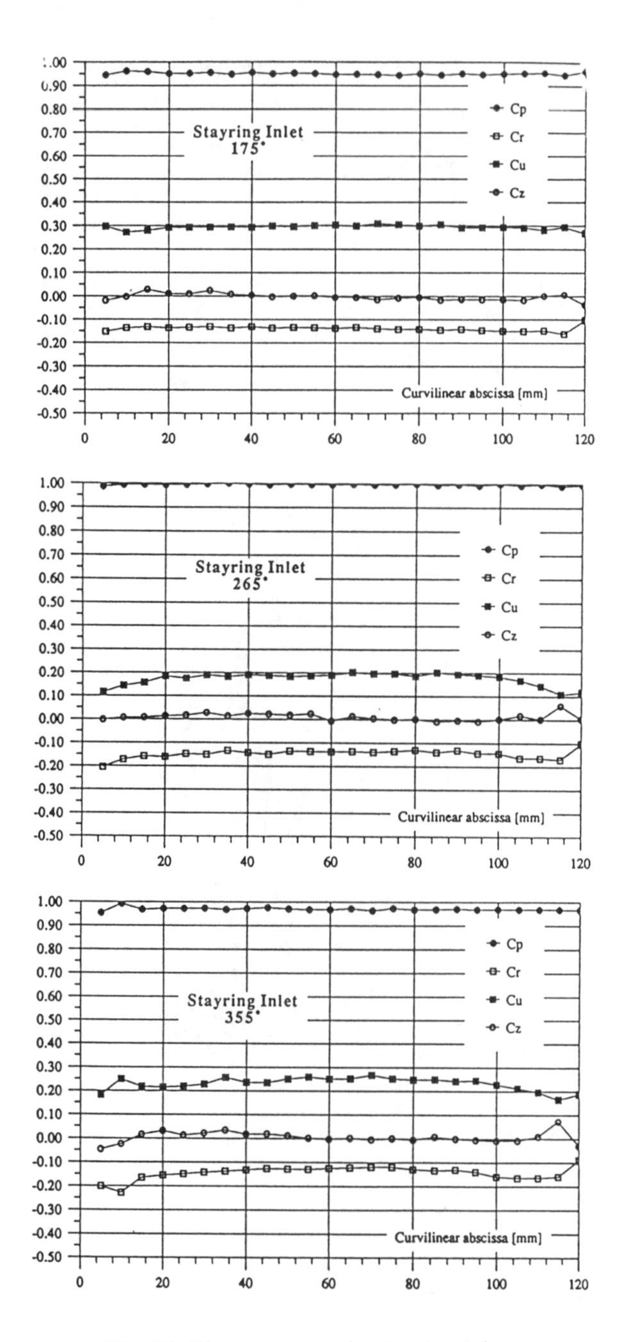

Fig. 16 Flow survey at the stayring inlet

These values are then related to the specific energy E provided by the test-rig measuring system in order to define the following pressure coefficient C_p of the nth transducer :

$$C_p = C_p^* \frac{p_I - p_{ref}}{\rho E} .$$

RESULTS

FLOW SURVEY

The operating point investigated in this study corresponds to the best efficiency point of the hillchart in Figure 7 : the flow coefficient $\varphi_{\bar{1}e_\wedge}$ and the energy coefficient ψ_{ref} are 0.286 and 1.072, respectively, and the guide-vane opening angle α is 25°.

Stayring inlet

Data corresponding to the flow survey at the stayring inlet is given in Figure 16. Static and total pressure coefficients C_p and C_{pt} and coefficients corresponding to the three flow velocity components are provided versus the traversing axis coordinates. Three sets of curves are given for a θ value of 175°, 265° and 355°, respectively. Owing to the influence of the fillet the flow is quasi two-dimensional, the c_z component being zero and the radial component remaining constant all along the periphery. A tangential component of the velocity coefficient c_u appears at these stations due to flow deviation by the stay vanes. Thus, the corresponding flow angle is very close to the stay vane angle ($\gamma_{ad} = 34°$).

Runner inlet

At the runner inlet no significant difference can be detected between the data obtained at different probe passages of this station. The mean flow is seen as axisymmetric by the pressure probe, thus the data reported in Figure 17 corresponds to the overall average of the measurements carried out at each probe passage. The tangential velocity component c_u, is uniform across the channel at the runner inlet. According to the channel geometry the meridional flow is deviated towards the vertical axis with an acceleration in the shroud region.

Cone section

The flow survey at the runner outlet shows a uniform meridional flow except in the mid core region, where the wake of the hub is felt ($R < 40$ mm), Figure 18. The positive value of the radial velocity component close to the wall of the cone ($R > 180$ mm) corresponds to the cone divergence. The linear outer distribution of the tangential component c_u is imposed by the trailing edge blade angle. Meanwhile a positive solid body rotation starts at the cone axis, extending up to $R = 20$ mm. The static pressure distribution is uniform except in the axis region where a lower pressure leads to predicting the onset of a whirl (see Figure 9).

Draft tube outlet

A set of probe passages were made at the draft tube outlet in order to perform a flow survey with a five hole probe. Unfortunately the large random flow instabilities did not allow to carry out precise velocity measurements with the probe. Nevertheless, wall pressure measurements were made at this section.

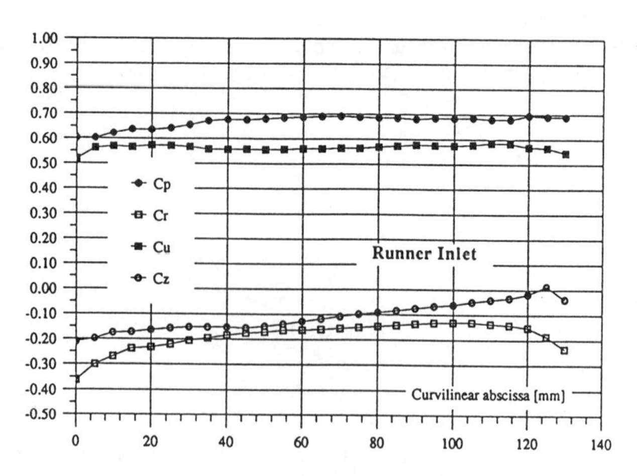

Fig. 17 Flow survey at the runner inlet

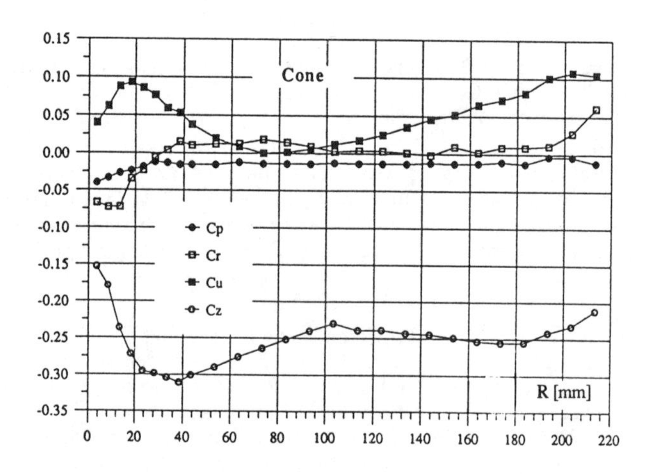

Fig. 18 Flow survey at the cone section

PRESSURE DISTRIBUTION IN THE RUNNER

The distribution of pressure coefficient along three theoretical streamlines close to, hub, shroud, and that corresponding to a midspan position is given in Figure 19, which represents only the second series of measurements in order to have a better readability. These values show that even at the best efficiency point the blade loading is far from optimal. There is low pressure distribution close to the shroud, leading to an inlet cavity development at the leading edge. Moreover, the strong adverse pressure gradient could lead to a possible flow separation.

The test was performed at a very high σ value ($\sigma \approx 1.5$), in order to have a pressure, higher than the atmospheric pressure, on the transducers. The specific hydraulic energy for this test was $E \approx 58$ [J·kg^{-1}].

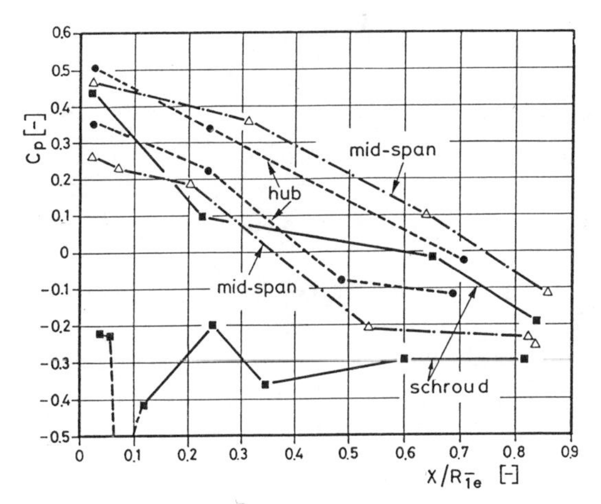

Fig. 19 Blade pressure

CONCLUSIONS

A complete experimental set-up has been developed in order to perform a flow study in a model of a Francis turbine. A five-hole probe mounted on a remote traversing system provides static pressure and velocity components along three stations in the machine. An original instrumentation is installed in the rotating runner in order to obtain the pressure distribution on the blades along three theoretical streamlines. This flow survey in the model provides an experimental set of data for CFD codes in order to check their ability to represent the flow in the complex geometry of a Francis turbine. Even though the accuracy of local velocity measurements can be improved, the experimental procedure provides an understanding of the flow conditions of hydraulic machines .

ACKNOWLEDGEMENT

The authors thanks all their colleagues and the technical staff of the IMHEF Test Rig.

REFERENCE

[1] HENRY, P. : *"Hydraulic machine model acceptance tests"*, Proceedings of International **Conference** on Hydropower, Water Power '85, Las Vegas, 1985, vol. 2, pp. 1258-1267.

NOTATIONS

Symbol		Description		Unit
C	:	Absolute velocity		$[m \cdot s^{-1}]$
Cr	:	Radial component of the absolute velocity		$[m \cdot s^{-1}]$
$Cz = -Ca$	:	Axial component of the absolute velocity		$[m \cdot s^{-1}]$
$Cu = C\theta$	:	Peripheral component of the absolute velocity		$[m \cdot s^{-1}]$
E	:	Specific hydraulic energy		$[J \cdot kg^{-1}]$
E_k	:	Specific kinetic energy		$[J \cdot kg^{-1}]$
E_{ref}	:	Specific hydraulic energy		$[J \cdot kg^{-1}]$
F	:	Calibration coefficient		[-]
G	:	Calibration coefficient		[-]
H	:	Calibration coefficient		[-]
H_s	:	Suction Head	$Z_{ref} - Z_{\bar{I}}$	[m]
I	:	Measuring section at the inlet of the spiral casing		[-]
$\bar{I}$	:	Measuring section at the outlet of the draft tube		[-]
L	:	Calibration coefficient		[-]
Q	:	Flow rate		$[m^3 \cdot s^{-1}]$
R	:	Radius		[m]
$R_{\bar{I}e}$	:	Reference radius at runner outlet		[m]
S	:	Surface		$[m^2]$
Z	:	Altitude		[m]
c	:	Velocity coefficient		[-]
g	:	Acceleration due to gravity		$[m \cdot s^{-2}]$
n_q	:	Specific speed		[-]
p	:	Static pressure		[Pa]
p_a	:	Atmospheric pressure		[Pa]
p_m	:	Average pressure		[-]
p_t	:	Total pressure		[Pa]
p_v	:	Vapour pressure		[Pa]

α	Guide vane opening		[°]
Φ	Yaw angle		[°]
γ_{do}	Guide vane closed position angle		[°]
η	Model turbine efficiency		[-]
$\varphi_{\bar{1}e}$	Discharge coefficient	$\dfrac{Q}{\pi\omega(R_{\bar{1}e})^3}$	[-]
ν	Specific speed	$\dfrac{\omega(Q/\pi)^{1/2}}{(2E)^{3/4}}$	[-]
θ	Pitch angle		[°]
θ_d	Guide vane angular position		[°]
θ_{ad}	Stay vane angular position		[°]
ρ	Water density		[$kg \cdot m^{-3}$]
σ	Thoma's cavitation number	$\dfrac{p_a - p_v - \rho g H_s}{\rho E}$	[-]
ω	Angular speed		[$rad \cdot s^{-1}$]
$\psi_{\bar{1}e}$	Energy coefficient	$\dfrac{2E}{\omega^2(R_{\bar{1}e})^2}$	[-]
ψ_{ref}	Energy coefficient	$\dfrac{2E_{ref}}{\omega^2(R_{\bar{1}e})^2}$	[-]

NORMALIZATION OF FLOW PROFILE DATA MEASURED AT RUNNER INLET

T. Kubota
Kanagawa University, Faculty of Mechanical Engineering
3-27-1, Rokkakubashi, Kanagawaku, Yokohama 221, Japan

INTRODUCTION

Measured flow profiles in a model turbine have inevitable measuring errors even with the most careful measuring technique. It is normal practice, therefore, that the measured data are normalized to conserve discharge, angular momentum and total specific hydraulic energy.

An Euler code can not include the effect of boundary layer developing on the wall. When the measured flow profile data are intended to be used as the reference inflow conditions for the Euler codes, the effect of boundary layers shall be removed from the measured profiles prior to the adjustment of the above normalizations.

A trial normalization including the removal of boundary layers is presented here for the flow profile data measured at the inlet of the GAMM runner, to analyze the runner with an Euler code.

ORIGINAL DATA ANALYSIS

The original radial and axial velocities c_r^* and c_z^* measured at the runner inlet are non-dimensionalized with the specific hydraulic energy E of the model turbine as follows:

$$c_r^* = \frac{C_r}{\sqrt{2E}} \tag{1}$$

$$c_z^* = \frac{C_z}{\sqrt{2E}} \tag{2}$$

and illustrated in Fig. 1 with solid marks and dashed lines versus the distance along a traverse line from the bottom ring ($b = 0$) to the head cover ($b = 1$). The distribution of radial and axial velocities is not uniform along the traverse line, and due to the curvature of the meridional contour of runner, both velocities generally increase at the runner band ($b = 0$) and decrease at the runner crown ($b = 1$) except in the boundary layers.

Similarly, the distribution of the original meridional and swirl velocities

$$c_m^* = \frac{C_m}{\sqrt{2E}} = \sqrt{c_r^{*2} + c_z^{*2}} \tag{3}$$

$$c_u^* = \frac{C_u}{\sqrt{2E}} \tag{4}$$

are shown in Fig. 2. The distribution of the meridional velocity c_m^* has the similar tendency as that of radial and axial velocities, however, the variation of the swirl velocity c_u^* is small except in the boundary layers.

The distributions of the original meridional flow angle α and swirl flow angle γ, i.e.

$$\alpha = \tan^{-1}\frac{c_z^*}{c_r^*} \tag{5}$$

$$\gamma = \tan^{-1}\frac{c_u^*}{c_m^*} \tag{6}$$

are illustrated in Fig. 3. The meridional flow angle α tends to direct radially inward at the crown and axially more inclined at the band. The swirl flow angle γ is circumferentially more inclined at the crown than at the band.

The correlation between the dynamic energy and the pressure energy for the original data can be confirmed with Fig. 4 in which

$$c^{*2} = c_r^{*2} + c_z^{*2} + c_u^{*2} \tag{7}$$

$$p^* = \frac{P - P_0}{\rho E} \tag{8}$$

where P_0 is the reference pressure for the pressure measurement. Band flow has more dynamic and less pressure energy; vice versa for the crown flow.

Finally, the total energy p_t^* and the Euler energy e^* of the original flow are shown in Fig. 5 in which

$$p_t^* = p^* + c^{*2} \tag{9}$$

$$e^* = \frac{E_{euler}}{E} = \frac{\omega R C_u}{E} \, . \tag{10}$$

The total energy p_t^* is more or less constant, though the scatter in the measured data is considerable. The Euler energy (flow circulation) e^* is higher at the band than at the crown.

EFFECT OF BOUNDARY LAYERS

Among the three velocity components of the original flow, the amount of swirl velocity c_u^* is the largest. Therefore, the effect of deceleration in the boundary layers due to friction becomes also large for the original swirl flow as shown in Fig. 2. This is the reason why the original radial velocity c_r^* is accelerated in the boundary layers as shown in Fig.1, corresponding to a reduction of centrifugal force. On the contrary, the original axial velocity c_z^* in the layer is a little decelerated as shown in Fig.1. Since the acceleration of c_r^* exceeds the deceleration of c_z^*, the meridional velocity c_m^* also accelerates in the layers (Fig.2), and the meridional flow angle α tends to decrease in the layers (Fig.3). The swirl flow angle γ in Fig. 3 and the Euler energy e^* in Fig. 5 for the original flow clearly show the effect of boundary layers near walls. There is no boundary layer distortion in the pressure energy distribution (Fig. 4) and within the measurement uncertainty in the total energy distribution (Fig. 5).

These effects of boundary layers shall be removed by extrapolating the tendency of the main flow profile into the boundary layer regions.

DISCHARGE CONTINUITY

The discharge that is computed by integrating the radial/axial velocities along a traverse line may usually differ from the discharge accurately measured by the flowmeter, mainly due to the non-axi-symmetric nature of the flow in the spiral case. The integrated discharge Q_{int} is

$$Q_{int} = 2\pi \int_{b=0}^{b=1} C_r R \, dZ + 2\pi \int_{b=0}^{b=1} C_z R \, dR \, . \tag{11}$$

This amounts to 0.339 m^3/s for the original velocities that is 8.9 % smaller than the measured discharge of 0.372 m^3/s.

The radial/axial and meridional velocities after the removal of boundary layer effects shall be normalized so as that the integrated discharge coincides with the measured discharge.

CONSERVATION OF EULER ENERGY

The mass averaged Euler energy at runner inlet $\overline{e^*}$

$$\overline{e^*} = \frac{\overline{E_{euler}}}{E} = \frac{\int_{b=0}^{b=1} \omega R \, C_u \, 2\pi R (C_r \, dZ + C_z \, dR)}{E Q} \tag{12}$$

can be checked with the energy e_P^* calculated from the measured turbine output P as follows:

$$e_P^* = \eta_h + \overline{e_2^*} = \frac{P}{\rho \left(Q - \Delta Q\right) E} + \overline{e_2^*} \tag{13}$$

where η_h is the hydraulic efficiency, $\overline{e_2^*}$ the mass averaged Euler energy at runner outlet, Q the turbine discharge, ΔQ the leakage discharge through the runner seals, respectively.

The η_h calculated from the original flow profile data has the following value:

$$\eta_{h,org} = \overline{e^*} - \overline{e_2^*} = 1.0607 - 0.0881 = 0.9726 .$$

which is too large compared with the measured efficiency of 0.920. The discrepancy $\Delta\eta_h$ between the calculated and measured efficiencies is to be shared by $\overline{e^*}$ and $\overline{e^*}_2$, and from Eq. (13) the value becomes

$$\overline{e^*} = 0.920 + (0.0881 + 0.0263) = 1.0344 .$$

The value is selected as a target for the normalization of the three velocity components, especially of the swirl velocity.

CONSERVATION OF TOTAL ENERGY

The specific hydraulic energy of the runner e_{th}^* can be expressed as follows:

$$e_{th}^* = \overline{p_t^*} - \overline{p_{t2}^*} \tag{14}$$

where $\overline{p_t^*}$ and $\overline{p_{t2}^*}$ are the mass averaged total energies at inlet and outlet of the runner calculated as follows:

$$\overline{p_t^*} = \frac{\int_{b=0}^{b=1} \left(p^* + c^{*2}\right) 2\pi R \left(C_r^* dZ + C_z^* \, dR\right)}{Q} . \tag{15}$$

The e_{th}^* calculated from the original flow profile data has the following value:

$$e_{th,org}^* = 1.0436 - 0.0556 = 0.9880$$

which is too large compared with the measured efficiency of 0.920.

The normalized total energy level shall be smaller than the original.

NORMALIZATION

Since no information is available for the theoretical main flow profile as well as the boundary layer growth along the walls, a trial and error approach shall be applied for the normalization of the original flow profile data measured at the GAMM runner inlet. The following criteria are applied for the normalization:

(a) The tendency of the main flow profile outside of boundary layers shall be maintained through the whole flow region as close as possible, and scatter in the data will be eliminated.

(b) The level of radial/axial and meridional velocities shall be increased to satisfy the discharge continuity.

(c) The level of swirl velocity and the Euler energy shall be reduced to meet the above mentioned target.

(d) The correlation between c_u^* and c_m^* shall be checked with the swirl flow angle γ, as well as the dynamic energy c^{*2}.

(e) The correlation among the dynamic energy c^{*2}, the pressure energy p^* and the total energy p_t^* shall be adjusted to keep the constant total energy along the traverse line.

A spread sheet of the Microsoft Excel for Macintosh is conveniently used with simultaneously displaying the Figs. 1 to 5 for the original and modified flow profiles. Common fluid engineering sense suggests the direction to modify the flow profile, and a few iterations are sufficient to obtain the normalized flow profile data that satisfy the criteria stated above. The normalized flow profiles are shown in Figs. 1 to 5 with blank marks and solid lines and are compared to the original profiles.

The normalized data have the following mass averaged energy:

$$e_{th,nor}^* = 1.0412 - 0.0556 = 0.9856$$

which is improved over the original both in value as well as in flatness, as shown in Fig. 5.

The flow profile data thus normalized are more appropriate to be applied as inflow conditions for the analysis of the GAMM runner with an Euler code.

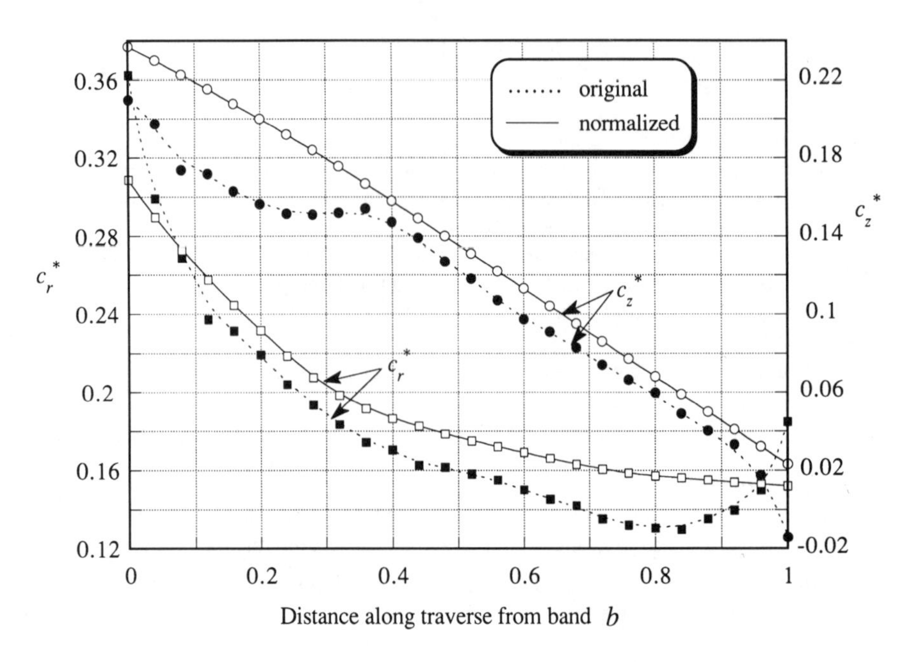

Fig. 1 Radial and axial velocity distributions

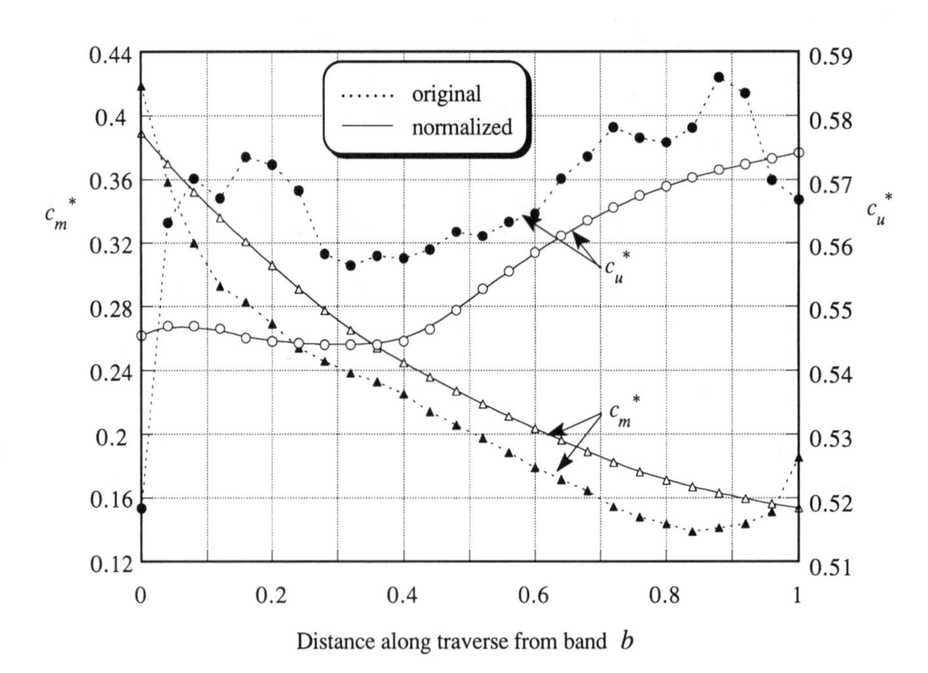

Fig. 2 Meridional and swirl velocity distributions

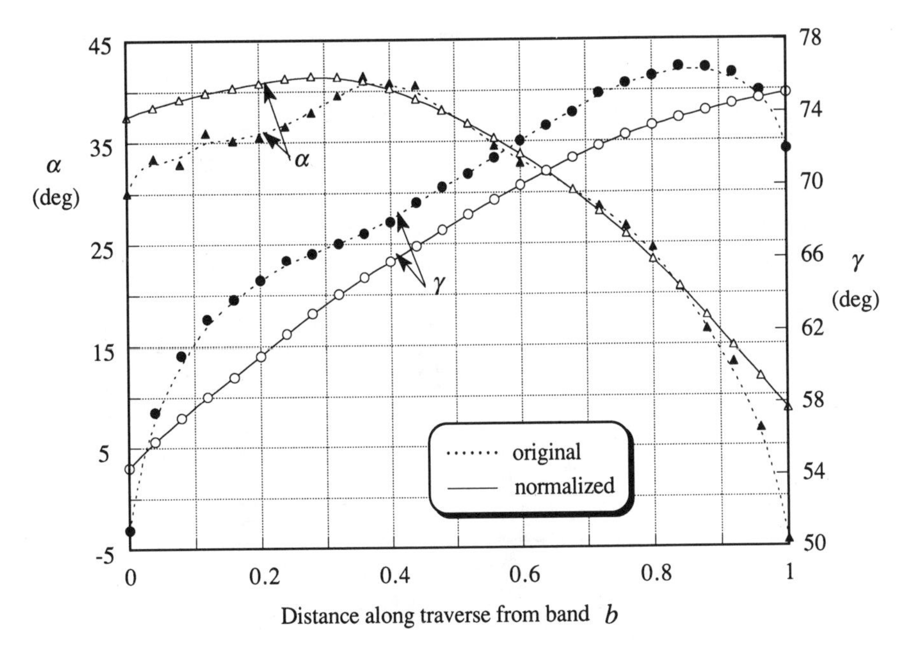

Fig. 3 Meridional and swirl flow angle distributions

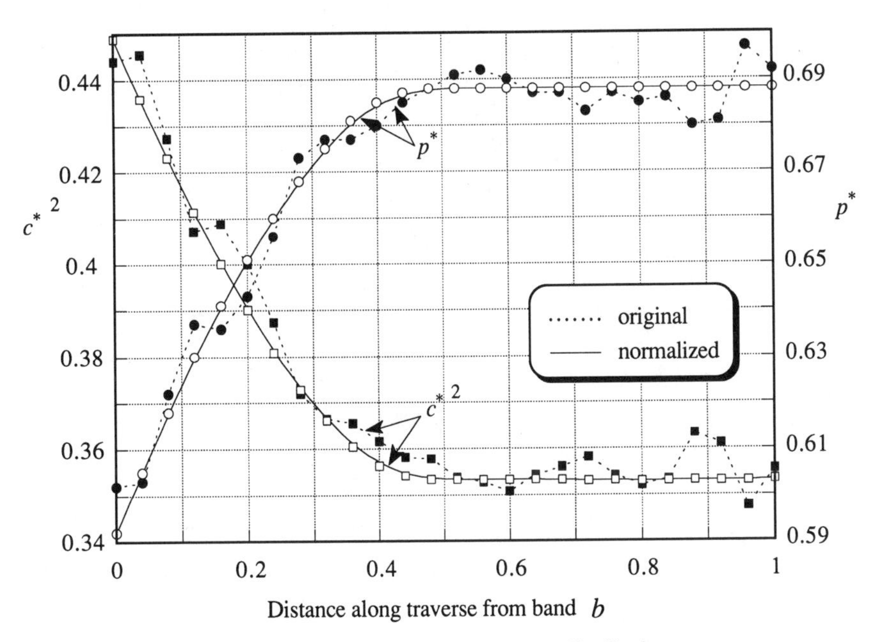

Fig. 4 Dynamic and pressure energy distributions

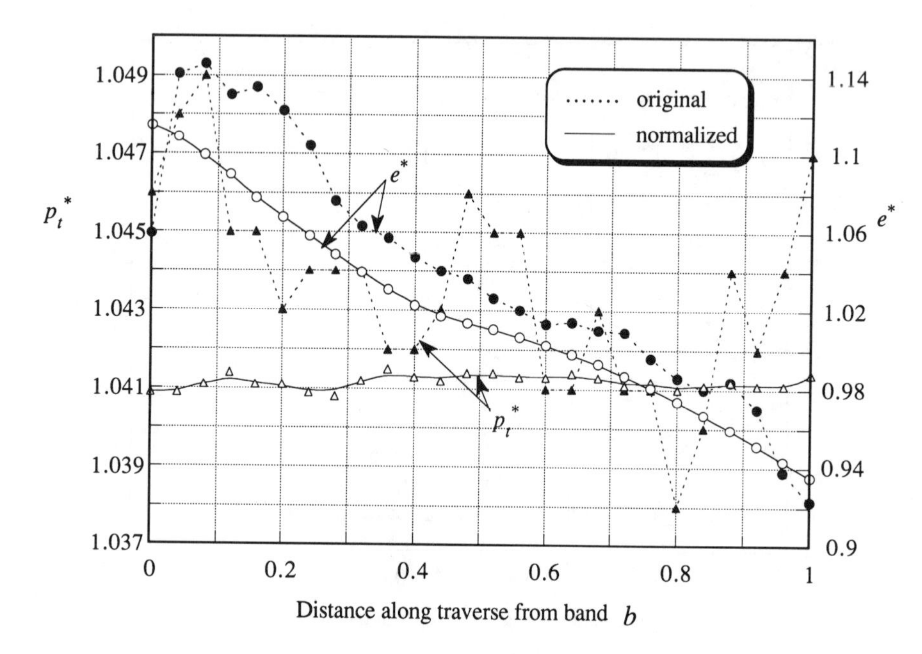

Fig. 5 Total and Euler energy distributions

Part 3

NUMERICAL FLOW SIMULATIONS

A 3D EULER SOLUTION OF FRANCIS RUNNER USING PSEUDO-COMPRESSIBILITY

C. Arakawa, M. Samejima
The University of Tokyo
7-3-1 Hongo, Bunkyo-ku, Tokyo 113, Japan

T. Kubota, R. Suzuki
Fuji Electric, Kawasaki
1-1 Tanabeshinden, Kawasaki-ku, 210 Kawasaki, Japan

SUMMARY

An implicit TVD three-dimensional Euler code has been extended by using the pseudo-compressibility for treating incompressible flows. The resulting code has been applied to the flow in the Francis runner for GAMM WORKSHOP 1989.

INTRODUCTION

One of the authors has already developed 3-D turbulent flow calculation code for advanced turboprop analysis (ATP), i.e., compressible flow calculation method for rotating blades[4], by using the two-equation turbulence model and the TVD scheme to capture shock waves clearly. In this paper we modify this code to calculate an incompressible flow, such as in a Francis runner, employing the pseudo-compressibility method which was introduced by Chorin[3]. In a first stage we describe here the results of inviscid flows. The purpose of this paper is not only to predict the flow accurately, but also to employ as computer an EWS(Engineering Work Station) which is now becoming popular and will become a personal computer in our design division in the near future.

GOVERNING EQUATIONS

Unsteady, inviscid, three-dimensional, incompressible flow with constant density is governed by the following equations written in rotating Cartesian coordinates. Using pseudo-compressibility, the continuity equation is modified by adding a time derivative term of the pressure which will tend to zero automatically as the steady state solution is obtained :

$$\frac{\partial Q}{\partial t} + \frac{\partial E}{\partial x} + \frac{\partial F}{\partial y} + \frac{\partial G}{\partial z} + H = 0 \qquad (1)$$

with

$$Q = \begin{bmatrix} p \\ u \\ v \\ w \end{bmatrix}, \quad E = \begin{bmatrix} \beta u \\ uu + p \\ uv \\ uw \end{bmatrix}, \quad F = \begin{bmatrix} \beta v \\ vu \\ vv + p \\ vw \end{bmatrix}, \quad G = \begin{bmatrix} \beta w \\ wu \\ wv \\ ww + p \end{bmatrix}, \quad H = \begin{bmatrix} 0 \\ -\Omega^2 x - 2\Omega v \\ -\Omega^2 y + 2\Omega u \\ 0 \end{bmatrix}.$$

Here, x,y, and z are Cartesian coordinates in the rotating frame; p,u,v, and w are pressure including the effects of the gravity force, and the components of the relative velocity, respectively. The factor β is a pseudo-compressibility parameter, and is fixed to 1 in this calculation. The direction of rotation is clockwise about the positive z direction and the angular velocity is Ω and all the variables are normalized properly.

General curvilinear coordinates (ξ, η, ζ) are introduced so as to accomodate complex three-dimensional geometries. After using the chain rule to transform the physical coordinates into curvilinear ones, we get the following form of the governing equations :

$$\frac{\partial \hat{Q}}{\partial \tau} + \frac{\partial \hat{E}}{\partial \xi} + \frac{\partial \hat{F}}{\partial \eta} + \frac{\partial \hat{G}}{\partial \zeta} + \hat{H} = 0 \tag{2}$$

with

$$\hat{Q} = J^{-1} \begin{bmatrix} p' \\ u \\ v \\ w \end{bmatrix}, \quad \hat{E} = J^{-1} \begin{bmatrix} \beta U \\ uU + \xi_x p' \\ vU + \xi_y p' \\ wU + \xi_z p' \end{bmatrix}, \quad \hat{F} = J^{-1} \begin{bmatrix} \beta V \\ uV + \eta_x p' \\ vV + \eta_y p' \\ wV + \eta_z p' \end{bmatrix}, \quad \hat{G} = J^{-1} \begin{bmatrix} \beta W \\ uW + \zeta_x p' \\ vW + \zeta_y p' \\ wW + \zeta_z p' \end{bmatrix},$$

$$\hat{H} = J^{-1} \begin{bmatrix} 0 \\ -2\Omega v \\ 2\Omega u \\ 0 \end{bmatrix}, \quad p' = p - \frac{1}{2}\Omega^2 (x'^2 + y'^2).$$

Here, U, V and W are contra-variant velocities; J is the Jacobian of the transformation.

NUMERICAL ALGORITHM

The numerical algorithm to advance eq.(2) in time is an implicit, approximately factored, finite-difference scheme, based on the method of Beam and Warming[1]. On the one hand for the spatial differencing of the explicit right-hand-side term, the upwind based high accuracy TVD scheme by Chakravarthy and Osher[2] is used, which has a good shock capturing property with no added artificial dissipation for the transonic flows and leads to minimum artificial dissipation even for incompressible flows. On the other hand, the implicit left-hand-side one relies on a diagonalized ADI method whereby the computational efficiency can be improved, because the steady state solution is indifferent to implicit operators. The spatial difference utilizes an upwind flux-split technique. Each ADI operator is decomposed into the product of lower and upper bidiagonal matrices by using the diagonally dominant factorization, and that operator is efficiently inverted by forward and backward sweeps[5]. The detailed information about this method is available in the references[1]-[2],[4]-[6].

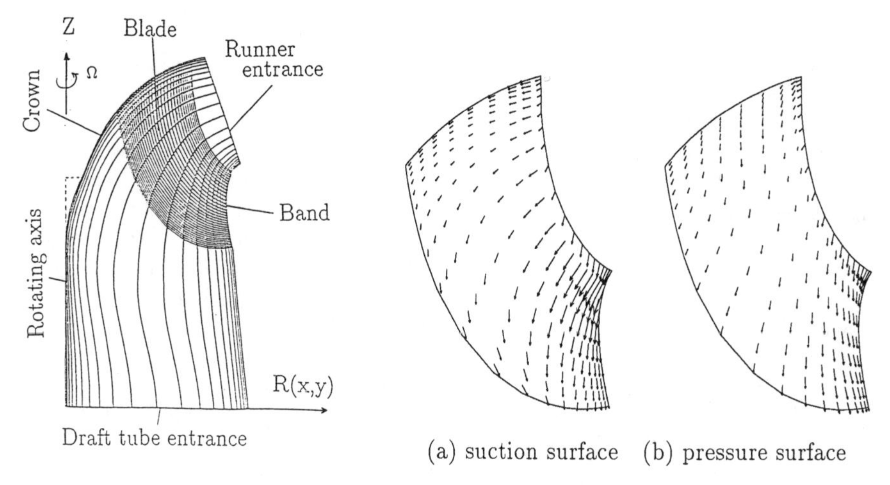

Fig. 1 Computational domain and the grid in Francis runner

Fig. 2 Velocity vectors on suction and pressure surface of Francis runner

GEOMETRY AND GRID GENERATION

The flow is assumed to be cyclic around the runner in steady-state conditions, so that the computational domain contains only one blade passage. Fig.1 shows the grid of ζ=const. plane mapped to the z-R plane. The in- and outflow boundary surfaces are specified by the two measurement cross-sections, and the crown surface is smoothly attached to the rotating axis for simplicity. Then all the boundary points are specified, and interior points are distributed by the algebraic interpolation. H-mesh topology is adopted on the blade-to-blade plane in order to avoid leading edge singularities and to facilitate the periodic boundary condition. Mesh clustering near the surfaces is imposed to enhance the numerical accuracy for the explicitly specified boundary conditions described below. The mesh we use in this calculation is 65x21x21.

BOUNDARY CONDITIONS

In this analysis, boundary conditions are all explicitly specified for simplicity. At the inflow boundary surface, all velocity components are given by the measured values, and the pressure is extrapolated from interior points. The axial symmetry requires that velocity components u and v are equal to 0 and that other values are given as averaged ones taken at surrounding points. On the crown, band and blade surfaces, impermeable conditions are specified as follows. The tangency condition is enforced by specifying the contra-variant velocity W=0 on the crown and band surfaces, and V=0 on the blade surfaces. Other contravariant velocities and the pressure are obtained by linear extrapolation from interior points. The pressure on the surfaces are calculated by solving the normal momentum equation. On the outflow boundary surface, the derivatives of all the quantities in the streamwise direction are assumed to be zero.

RESULTS AND DISCUSSIONS

The calculated flow pattern expressed in the velocity vector form seems reasonable at the design condition, because the water flows approximately in the direction of η=const. on the suction side, and it flows from the crown to the band in the region near the leading edge on the pressure side (Fig.2). The pressure contours on both of the pressure and suction surfaces are shown in Fig.3. We find here a strong suction peak near the corner of leading edge and band on the suction surface. As it turned out later, the mass-flow rate at the runner inlet obtained by integrating the velocity distribution conflicts with the measured value. This is why our code predicts larger tangential velocity at the middle measurement line and relatively high pressure level on the pressure side of the runner. For the better prediction, some modification must be made in the velocity distributions at the runner inlet. The non-dimensional torque is calculated 0.24960 and the efficiency is predicted to be 88.3% according to the formulation proposed by the organizer.

This calculation is carried out using the super-EWS computer with the vector processor whose speed is 16MIPS. The CPU time to get the converged result is about 6 hours and a half for 1000 iterations, which is acceptable for engineering purposes. Table 1 shows the computer related quantities and Fig. 4 shows the convergence history.

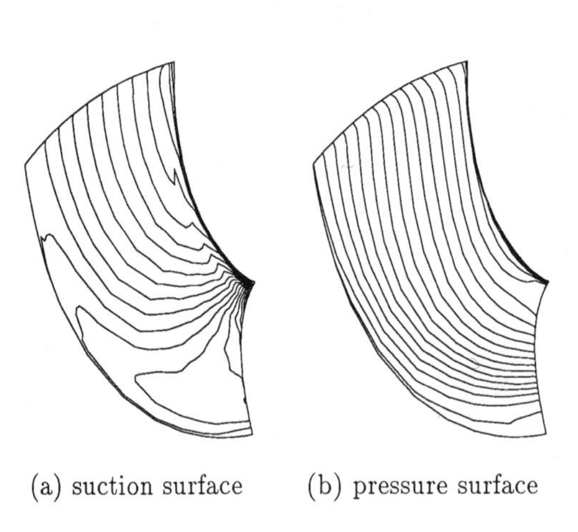

(a) suction surface (b) pressure surface

Fig. 3 Pressure contours on suction and pressure surface of Francis runner

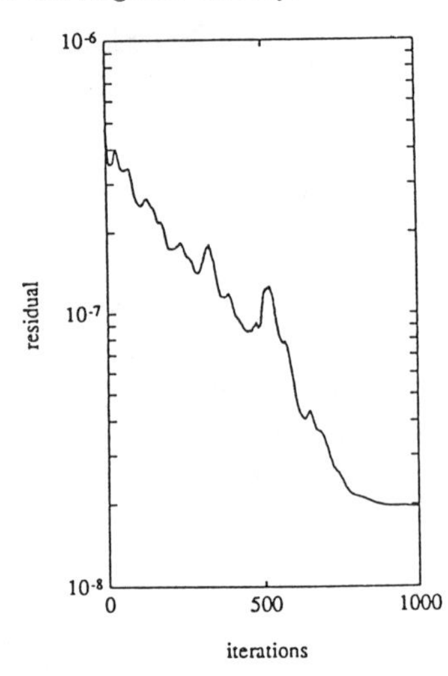

Fig. 4 Convergence history.

CONCLUSIONS

A three-dimensional Euler code for incompressible calculations using the artificial compressibility concept, the high accuracy TVD scheme and the implicit formulation has been extended and applied to analyze the flow in the Francis runner. The performances and flow pattern derived in the calculation here are in reasonable agreement with the experiment. However, the runner inlet velocity must be modified for the conservation of the mass flow to predict the pressure levels on the runner surfaces more accurately.

Table 1   Computer related quantities

computer type	TITAN 1500 (1 cpu)
number of grid points used	28665($65 \times 21 \times 21$)
cpu time	22932s(6:22:12)
cpu time/iteration	22.973s
(cpu time)/(number of grid cells)	0.89578s
(cpu time)/(iteration*number of grid cells)	0.00089578s
ratio of vectorization	2.3421
parallelization speedup	No parallelization

ACKNOWLEDGEMENT

The authors wish to thank Dr. Y. Matsuo and Dr. S. Saito in National Aerospace Laboratory, Japan for many helpful discussions.

REFERENCES

[1] BEAM, R.M., WARMING, R.F. : "An implicit finite-difference scheme for hyperbolic systems in conservation form", J. Computational Physics **22** (1976) pp. 87-110.

[2] CHAKRAVARTHY, S.R., OSCHER, S. : "A new class of high accuracy TVD schemes for hyperbolic conservation laws", AIAA Paper 85 - 0363 (1985).

[3] CHORIN, A.J. : "A numerical method for solving incompressible flow problems", J. Comp. Phys. **2** (1967) pp. 12-26.

[4] MATSUO, Y., ARAKAWA, C., SAITO, S., KOBAYASHI, H. : "Navier-Stokes simulations around a propfan using higher-order upwind schemes", AIAA Paper 89 - 2699 (1989).

[5] OBAYASHI, S., FUJII, K. : "Computation of three -dimensional viscous transonic flows with the LU factored scheme", AIAA Paper 85 - 1510 (1985).

[6] ROGERS, S.E., KWAK, D., KIRIS, C. : "Numerical solutions of the incompressible Navier-Stokes equations for steady-state and time-dependant problems", AIAA Paper 89-0463 (1989).

NUMERICAL INVISCID FLOW ANALYSIS OF THE GAMM FRANCIS RUNNER

J.T. Billdal, Ø. Jacobsen
SINTEF Fluid Machinery
7034 TRONDHEIM, NORWAY

K. Bratsberg
Kværner Brug A/S
Postboks 3610, Gamlebyen, 0135 OSLO 1, NORWAY

H.I. Andersson, H. Brekke
The Norwegian Institute of Technology
7034 TRONDHEIM, NORWAY

SUMMARY

The steady incompressible flow in the FRANCIS RUNNER is numerically computed by solving the three-dimensional Euler equations on an H-grid. A finite volume discretization scheme with an explicit time integration and absorbing inflow/outflow boundary conditions is used. By comparing the numerical solution with measurements a good agreement was obtained.

INTRODUCTION

SINTEF (The Foundation for Scientific and Industrial Research at the Norwegian Institute of Technology) has been engaged in the analysis of flows in hydraulic machinery for nearly a decade. With the installation of the CRAY X-MP/28 supercomputer in Trondheim in 1987, the otherwise time-consuming three-dimensional solutions of the Euler and Navier-Stokes equations have become tractable. In a collaboration between SINTEF, The Norwegian Institute of Technology (NTH) and the local industry, a continuous development of computer programs is taking place. This development includes a strive for more efficient algorithms as well as more realistic model equations, which includes viscous, turbulent and unsteady effects. The present paper describes the numerical method used to solve the three-dimensional, steady, incompressible Euler equations governing the inviscid flow in a Francis turbine.

THE FINITE VOLUME METHOD

To solve the steady, incompressible Euler equations in hydraulic machinery components we adopted the artificial compressibility concept in combination with a finite-volume formulation of the governing equations. Chorin's artificial compressibility approach [2] was originally applied to solve the incompressible Navier-Stokes equations by finite-difference techniques [2, 7], but has also been used to solve finite-volume formulations of the inviscid Euler equations [5, 8]. More recently, centrifugal and Coriolis terms have been added to the momentum equations, in order to simulate flows in a rotating coordinate system [3, 6, 9, 10].

Basic principles

A steady solution of the 3D Euler equations is obtained by marching in pseudo-time from some initial flow field until an asymptotic state is achieved. The method is based on a cell-centered finite-volume scheme using centered second-order-accurate space discretization with explicit three-stage Runge-Kutta time stepping [1]. Convergence to steady state is accelerated by using the concept of variable time stepping. To avoid nonlinear instabilities (saw-tooth waves) a constant 4th order dissipative term is added to the scheme throughout the domain.

Finite-volume mesh

An algebraic technique called "transfinite interpolation" has been used to generate an H-type grid for the runner. The flow field was discretized by 78x15x20 grid points. Figure 1 shows the grid in a meridional plane while Figure 2 shows it between the blades.

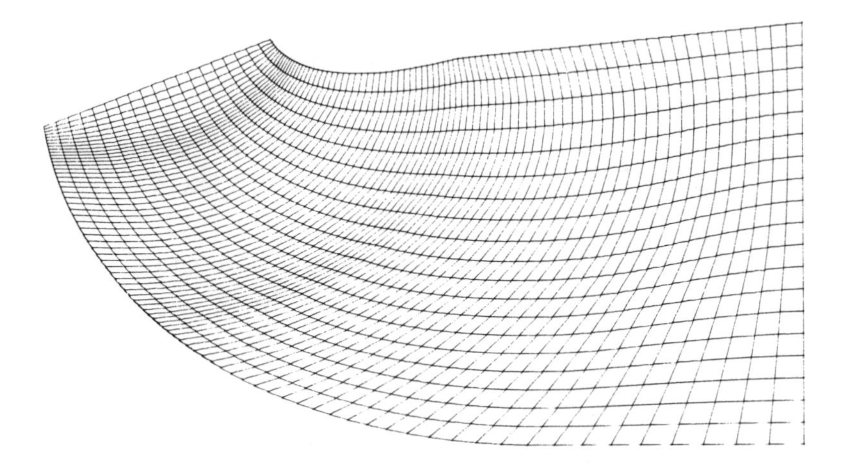

Figure 1 Meridional grid surface.

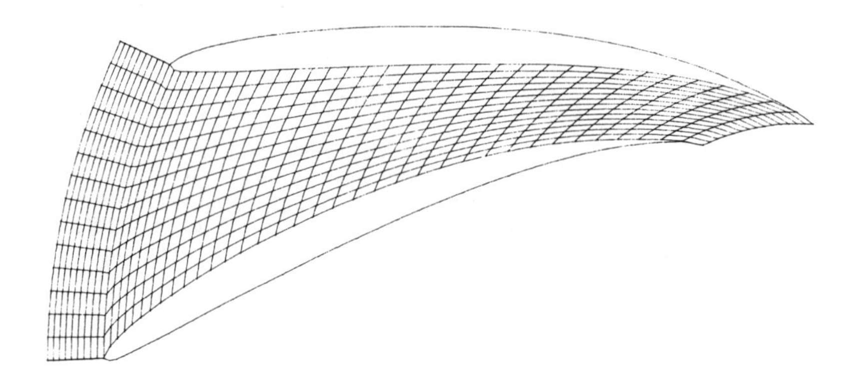

Figure 2 The grid at crown between the blades

Boundary conditions

At inflow/outflow boundaries the theory of absorbing boundary conditions [4] in its simplest form is applied. This is done by linearizing the governing equations locally and computing the characteristic variables in the normal direction. Those characteristic variables, which are advected into the domain, are then fixed to the desired values, whereas those which are advected out of the domain are linearly extrapolated from the interior to the boundary. The resulting complete set of characteristic variables is then transformed back to physical variables and used to compute the desired fluxes. It can be shown that this technique gives stable boundary conditions and absorbs waves very effectively.

At solid walls the mass flux is zero but the pressure contributes to the momentum flux and is found by a linear extrapolation from the interior domain. The periodicity condition imposed along periodic boundaries is such that these cells are updated just as interior cells.

Computer related quantities

- computer type: CRAY X-MP/28
- number of grid points: 23.400
- cpu time: 468 sec.
- (cpu time)/iterations: 0.3 sec.
- (cpu time)/(number of grid cells): 0.0228 sec.
- (cpu time)/(iterations * number of grid cells): 1.46 E-5 sec.
- convergence criterion: $\max [\log ((P/\rho)^{n+1} - (P/\rho)^{n})] \leq -3.5$

The convergence history is shown in Figure 3.

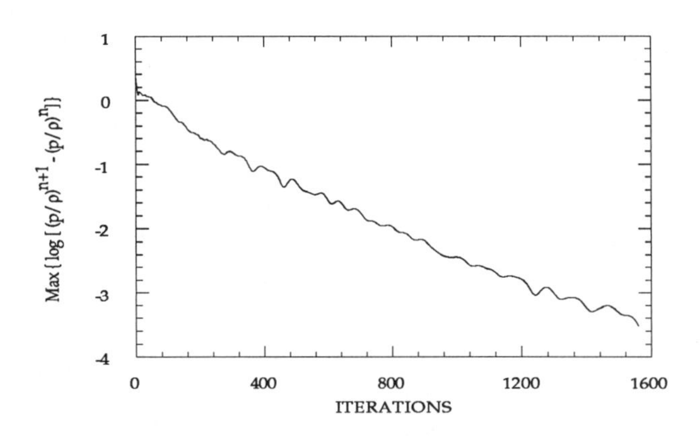

Figure 3 The convergence history

COMPUTATIONAL RESULTS AND DISCUSSIONS

By using absorbing boundary conditions, as outlined above, the inflow is allowed to adjust itself to fulfil the requirement of global continuity, i.e. $Q_1 = Q_2 = 0.377\ m^3/sec$. The correct upstream influence of the stagnation pressure on the flow conditions at the inlet was thus reproduced by the simulation. In our opinion this detail is very important, since it seems obvious that the forcing of a uniform rotational-symmetric flow into the blade cascade of the runner close to the inlet inevitably leads to inaccurate velocity and pressure distributions.

The computational results

By analysing the isobars in the blade to blade stream surface (S2) close to the crown and in the middle section, a smooth pressure distribution increasing from suction to pressure side is observed in Figure 4. In the section close to the band, however, the predicted pressure close to the blade inlet on the suction side is approximately 1 m lower than the draft tube pressure. This is most likely caused by the blade leaning angle, which may be changed to increase the pressure at the band. Changing the number of the blades or their angles or profiles may also be useful means to improve the pressure distribution.

The velocity vectors along the blades' pressure and suction sides are shown in Figure 5. A strong cross-flow along the leading edge from crown to band is observed, together with a tendency of flow from the pressure to suction side around the inlet edge. The pressure distribution on the blades' suction side shows a decreasing pressure towards the band, which leads to a starting inlet cavitation at this point if the draft tube pressure is decreased to σ_{cr}.

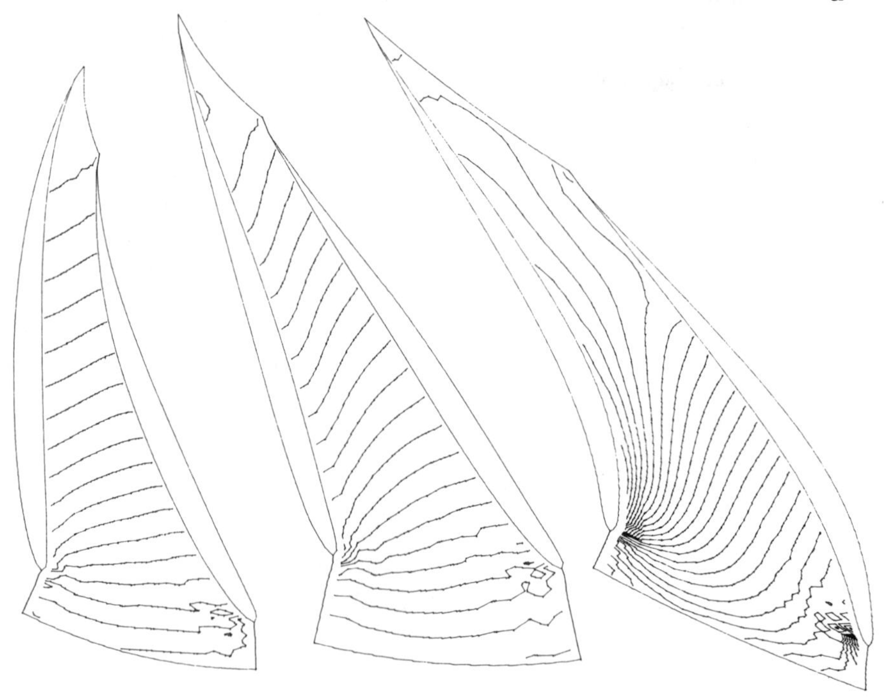

Figure 4 Isobars. Equidistance 0.2 m.
Crown (left), middle (middle), band (right).

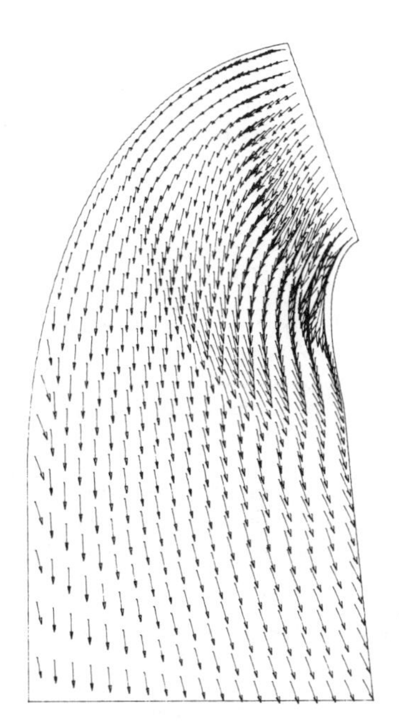

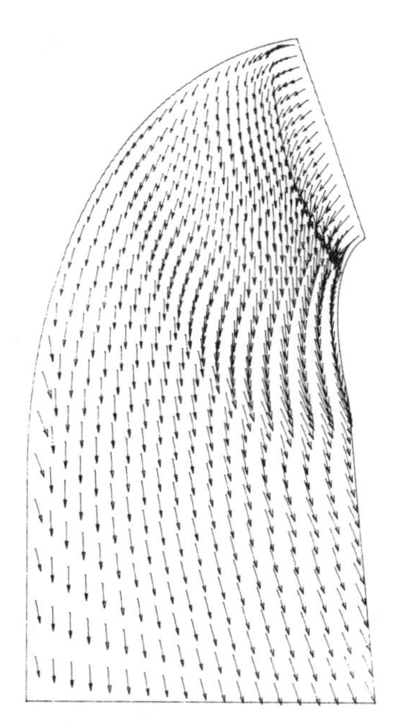

Figure 5 Velocity vector plots.
Suction side (left) and pressure side (right).

Comparison of computation and measurements

Comparisons of predicted and measured data are provided by the organizers of the GAMM-Workshop in a separate section of this proceedings. An expected agreement is observed by comparing measured and computed mean values of the pressure and velocity components at the **Inlet** and **Outlet**. A remarkable good agreement can be observed at **Runner-Middle**. This is very encouraging since **Runner-Middle** is away from the boundaries of the calculation domain (at which the boundary conditions were imposed).

By considering the results along the blades' suction and pressure sides, some deviations between the computed and measured (averaged quantities) data are observed. The results are satisfactory, however, except in the vicinity of the extremely low pressure peak at the blade inlet in Section 15. It is interesting to notice that a more pronounced peak pressure drop was obtained in a very recent computation with a somewhat refined computational grid (not presented during the Lausanne-Workshop).

CONCLUSIONS

By comparing the computations based on Euler equations with the measurements a good agreement was obtained. However, improved results have already been obtained by a time-averaged Navier-Stokes solver with a turbulence model. Current work at SINTEF/NTH also includes development of a time dependent code for comparison of dynamic measurements of the flow in runners. (The non-stationary flow in the runner is caused by the wakes from the guide vanes leading to a loss in efficiency). The results from these more advanced studies will be presented in forthcoming papers.

ACKNOWLEDGEMENT

This work has been sponsored by The Royal Norwegian Council for Scientific and Industrial Research (NTNF), The Norwegian State Power Board and Kværner Brug A/S.

REFERENCES

[1] BILLDAL, J.T.: "A finite-volume solution technique for the Navier-Stokes equations governing incompressible flow", SINTEF Report no. STF67 A 89003, 1989.

[2] CHORIN, A.J.: "A numerical method for solving incompressible viscous flow problems", J. Comp. Phys., 2 (1967) pp. 12-26.

[3] ELIASSON, P., KROUTHEN, B.: "The construction of an incompressible inviscid Euler solver for rotating systems and a comparative study of two Euler pump flow solutions", Proc. 1st Scandinavian Conference on Viscous and Turbulent Flow in Hydraulic Machinery, Trondheim, April 1987, pp. B1-B9. Also: FFA TN 1987-03.

[4] ENGQUIST, B., MAJDA, A.: "Absorbing boundary conditions for the numerical simulation of waves", Math. Comp., 31 (1977) pp. 629-651.

[5] ERIKSSON, L.E., RIZZI, A., THERRE, J.P.: "Numerical solutions of the steady incompressible Euler equations applied to water turbines", AIAA paper 84-2145 (1984).

[6] GOEDE, E., CUENOD, R., BACHMANN, P.: "Theoretical and experimental investigation of the flow field around a Francis runner", Proc. 14th IAHR Symposium, Trondheim (1988) pp. 503-514.

[7] PEYRET, R., TAYLOR, T.D.: "Computational methods for fluid flow", Springer Verlag, New York, 1983.

[8] RIZZI, A., ERIKSSON, L.E.: "Computation of inviscid flow with rotation", J. Fluid Mech., 153 (1985) pp. 275-312.

[9] SAXER, A., FELICI, H., NEURY, C., RYHMING, I.L.: "Euler flows in hydraulic turbines and ducts related to boundary conditions formulation", Proc. 7th GAMM Conference on Numerical Methods in Fluid Mechanics, Louvain-la-Neuve, September 1987.

[10] THIBAUD, F., DROTZ, A., SOTTAS, G.: "Validation of an Euler code for hydraulic turbines", AGARD Symposium on Validation of Computational Fluid Dynamics, Lisbon, May 1988, pp. 27.1-27.14.

EULER SIMULATIONS OF FLOW IN A FRANCIS DISTRIBUTOR AND RUNNER

A. Bottaro, A. Drotz, P. Gamba, G. Sottas
Swiss Federal Institute of Technology (EPFL)
Hydraulic Machines and Fluid Mechanics Institute (IMHEF)
ME - Ecublens, CH - 1015 Lausanne, Switzerland

C. Neury
Ateliers de Constructions Mécaniques de Vevey S.A. (ACMV)
CH - 1800 Vevey, Switzerland

SUMMARY

The inviscid rotational flow in the distributor and runner of a Francis turbine has been computed by simulation of the three-dimensional Euler equations. A finite volume technique is used for the discretization of the governing equations, which are modified via the artificial compressibility procedure, and steady solutions are obtained by explicit (pseudo-)time marching. For both the configurations examined good agreement with experimental data is observed.

INTRODUCTION

The flow in a Francis water turbine can be successfully analyzed by using the Euler equations [6-8]. They admit transport of vorticity which makes them more attractive than simpler potential techniques. The approximation involved in neglecting viscosity is acceptable because of the high relative velocities in the turbine and favorable pressure gradients, and the consequent small thickness of the boundary layer on the blades when we are close to optimum design parameters. It will be shown that Euler calculations estimate well the pressure load on the blades, and pinpoint correctly the areas of high and low pressure. Low pressure regions in the blades of Francis runners can be associated with cavitation phenomena. A study taking into account all the wealth of physical effects in a Francis installation (two-phase flow, turbulence, ...) is not currently feasible.

NUMERICAL TECHNIQUE

A convenient and robust way to obtain a steady state solution from the equations of motion of an incompressible fluid is the artificial compressibility technique [2]. The method is such that the system of continuity and Euler equations, in conservative form, is made hyperbolic by the addition of a pressure time-derivative to the equation expressing mass conservation. By doing this an explicit pressure equation is introduced, and the complete system can be dealt with through one of the many well developed techniques for hyperbolic systems. Elliptic solvers are somewhat less developed. The technique used has, however, the shortcoming that only the steady state solution is valid, and the pseudo-transient has to be regarded only as an iteration procedure without physical significance.

The equations are discretized by a finite volume procedure, that employs an array of non-staggered non-overlapping grid cells to cover the calculation domain. Convective fluxes are treated with a central difference discretization and an explicit fourth order artificial dissipation term is added to the scheme to damp dispersion errors. Care must be taken to prevent this artificial viscosity term to act as a source of vorticity generation; this problem, which is very delicate, is linked to the smoothness of the grid. An explicit four-stage Runge-Kutta scheme is adopted to march in time [3]. It is "only" first order accurate, but it allows a CFL limit on the time step higher than other Runge-Kutta schemes, like the one described in [6], and is such that steady state is reached faster. Local time stepping is adopted to accelerate convergence.

The boundary conditions and their implementation are of considerable importance. At the open boundaries (inlet and outlet), conditions based on the theory of characteristics have been adopted [7]. The number of conditions to impose depends on the number of incoming characteristics at the surface. At an inlet the three velocity components are specified, while at an outlet we specify the pressure; the other variables are extrapolated from the interior. Although this is a conventional procedure, the choice made is not unique, and different variables can be specified at the open surface. The values used come from measurements. Problems arise if at the open surface there is recirculation of fluid; this may be the case at certain off-design point operating conditions at the outlet of Francis runners, where vortex breakdown phenomena take place, and at the outlet of draft tubes, when flow separation occurs. In these instances there is fluid crossing the boundary in two opposite directions, and either one or three variables have to be specified on the same surface depending on the direction of the flow. A technique that does not appeal to the characteristics may be preferable in these cases [1]. At periodic boundaries Dirichlet conditions are adopted, which are such that a periodic surface is updated just as an interior surface. At solid walls free slip conditions are imposed on the velocity, while the pressure is extrapolated to the wall from neighboring control volumes.

A considerable amount of details on the technique adopted is provided in [6].

The validation of the code is described in [8]

MESH GENERATION

For each of the two geometries, the distributor and the runner, the flow simulation has been performed on a portion of the whole domain, extracted from the whole by exploiting the fact that there is azimuthal periodicity. The three dimensional H-H grids used for the computations are shown in Figs. 1 and 2. For technical reasons, the geometries we have generated are such that the runner rotates in the *clockwise* sense, contrary to the specifications.

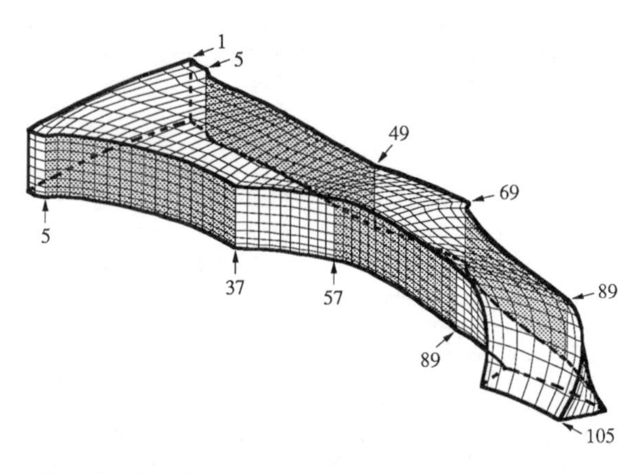

Fig. 1. Interblade domain with grid used for the flow simulation in the distributor (the blades are shown shaded).

The distributor is composed of 105 points in

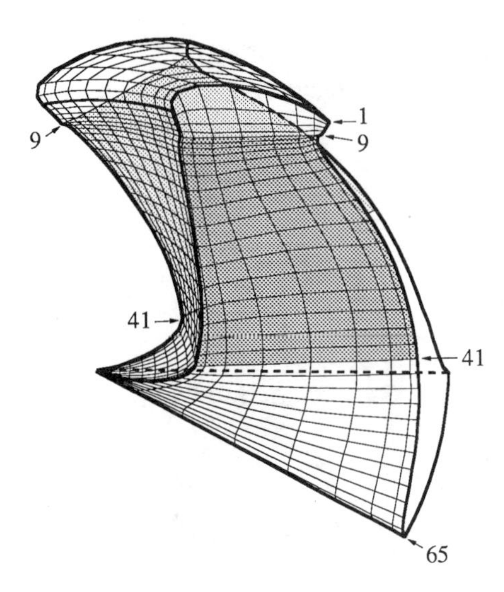

Fig. 2. Interblade domain with grid used for the flow simulation in the runner (the blades are shown shaded).

the streamwise direction and of 17x17 points in the crosswise directions. The runner employs 65 nodes in the streamwise direction, 21 along the azimuthal direction and 17 along the radius.

The meshes have been generated by an algebraic technique (transfinite interpolation) documented in [4]. It allows for the possibility of having quasi-orthogonal grid lines by direct specification of the normal derivatives (of arbitrary order) at the boundaries. Such a form of mesh control has been used only for the distributor. It should be remarked that, in the case of the distributor, there is a "shifting" of streamwise mesh points in the region between stay vane and guide vane at either side in the azimuthal direction to get volumes with almost orthogonal angles. The *number* of nodes in such a periodic window is, however, the same along both azimuthal boundaries (see Fig. 1).

Converged solutions have been obtained in 3 steps : the simulation was done first on a coarser grid extracted out of a finer one by dividing the number of cells in the three parametric directions by four; subsequently the coarse-grid output is used as an initial condition for a medium-mesh run, and similarly for the finest mesh. This procedure, which involves tri-linear interpolation once convergence on a coarser mesh is achieved, accelerates convergence on the finer mesh [8].

RESULTS

DISTRIBUTOR

The results for the distributor compare well with experiments, as detailed in the general paper of these proceedings. Differences may be ascribed to numerical perturbations, travelling back and forth in the domain, arising from the fact that the inlet (where pressure and velocity probes have been placed) is very close to the leading edge of the blades. In addition to that, only one vertical axis has been probed within the computational domain, so that numerical inlet conditions have been generated by assuming axisymmetry. By doing so there is a difference between numerical and experimental flow rate of about ten per cent.

From a qualitative viewpoint it should be remarked that three dimensional graphical representations provide valuable information on the nature of the flow and pressure fields. In particular, as evidenced in Fig. 3, the

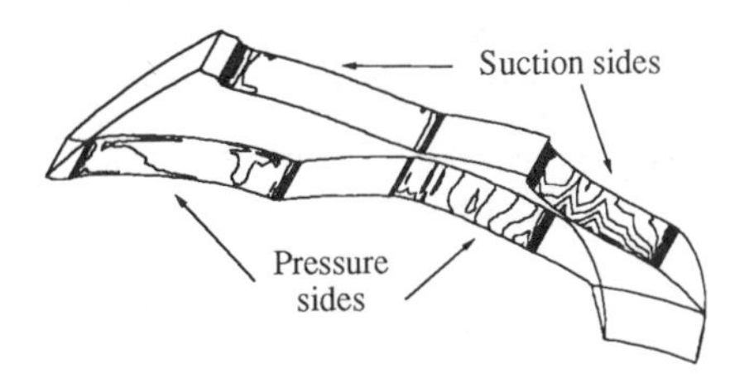

Fig. 3. Isobars on pressure and suction sides of the stay and guide vanes of the distributor.

pressure varies irregularly on the pressure side of the stay vane, while a quite regular streamwise decrease appears on the guide vane. Steep pressure gradients close to the leading edge of the guide vane should partly be ascribed to the mesh design. Pressure variations on the guide vane appear, again, more regular, while a small vortex, developing near the leading edge of the stay vane has been noticed. This could be due to the artificial dissipation term employed. The pressure distribution on the guide vane blades provides some clue on the acceleration the fluid undergoes in the second half of the distributor. The fluid motion can be well illustrated by means of particle traces. In Fig. 4 one can see the streamwise development of three sets of sixteen particles released at the entrance on three different horizontal planes. While the particles on the middle plane tend to remain on the plane itself, those on the two other planes near the top and bottom of the distributor are considerably deflected.

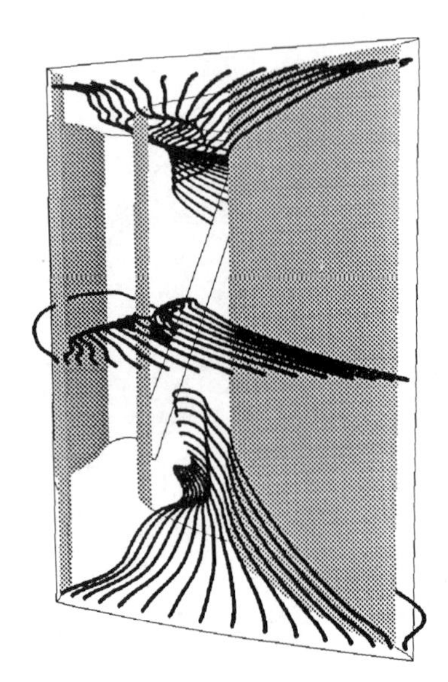

Fig. 4. Particle traces in the distributor.

RUNNER

Two different series of calculations have been performed for the Francis runner. The reason for this lies in the fact that, using the experimental data provided by the organizer at the entrance of the rotating part in an axisymmetric fashion (to provide an inlet condition), causes the mass flow rate to be inferior to the measured one by about 7%. Therefore, we denote as *case 1* the calculations performed with the data provided, and as *case 2* the computations done after having adjusted the inlet velocity distribution to satisfy the mass balance specified. It is clear that by adopting the second procedure we modify the inlet flow angles, with consequences on the flow distribution in the runner and on the efficiency.

The results obtained in both cases are in good agreement with measured values, as discussed elsewhere. The reasons for the differences found are the same as those described at the beginning of the *DISTRIBUTOR* section.

In Fig. 5 the pressure distribution on suction and pressure side of the runner is shown for *case 2*. We can pinpoint a minimum in pressure on the lower part of the suction side between leading edge and band, which agrees with the experimentally observed region of cavitation. Wiggles in the isolines are almost absent. On the pressure side the pressure is seen to vary very regularly, as expected. Results are almost identical in *case 1*. Differences among the two cases arise when inspecting velocity vectors on the blades. In Fig. 6 we have the velocity field on suction and pressure sides for *case 2*. The two fields look very much alike, and are both quite typical of high pressure blades; on both sides we remark regions of recirculation near the entrance. This is dramatically different from the result obtained for *case 1* on the blade's suction side, shown in Fig. 7.

Sixteen particles have been released at the entrance of the runner on a vertical line, and followed in their paths (Fig. 8). The traces separate from one another as they encounter the leading edge of the blade, and are approximately equally distributed between those "choosing" to follow the blade on the low pressure region, and those "choosing" the high pressure region. About halfway through the length of the leading edge there is evidence for the appearence of a leading edge vortex. Trajectories eventually tend to merge and to swirl out of the runner to the draft tube.

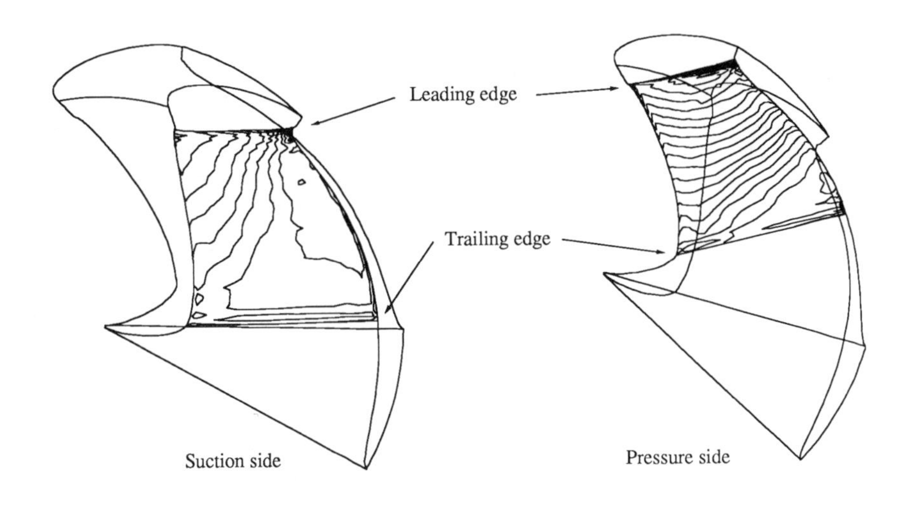

Fig. 5. Isobars on suction and pressure sides of the runner blades, case 2.

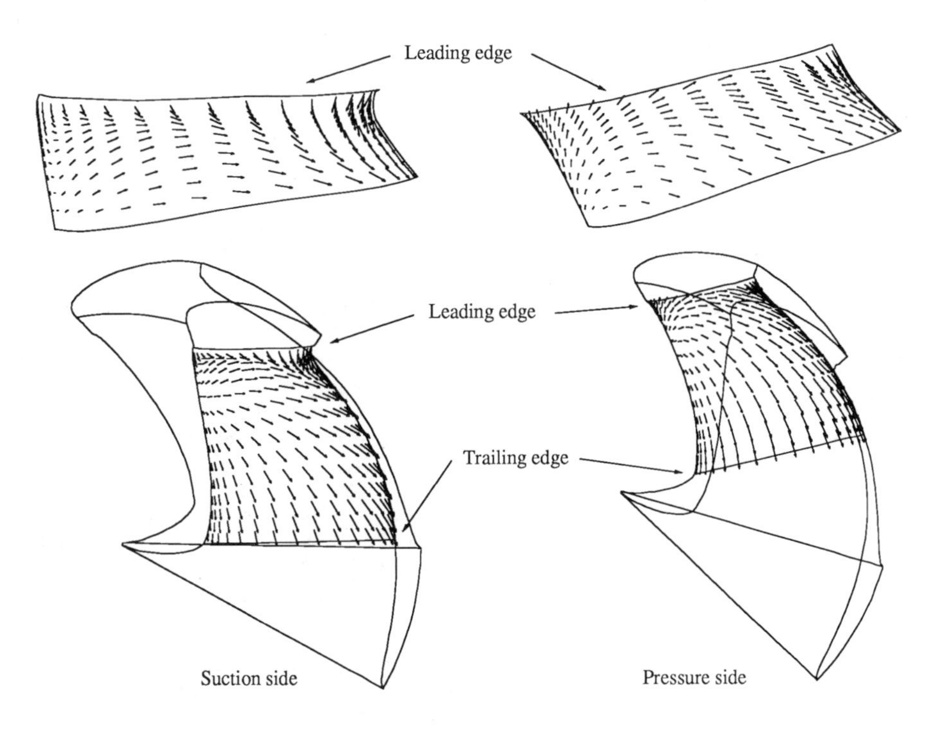

Fig. 6. Velocity fields on suction and pressure sides of the runner blades, case 2. Enlargements of the leading edge regions are shown above the complete interblade volume for each side.

Table 1. Computer related quantities for the distributor and the runner

	DISTRIBUTOR	RUNNER
Computer type	Cray 2 Program run on one processor	Cray 2 Program run on one processor
Number of grid points	30345	23205
CPU time	3198 [s] for 13500 time steps, to bring the average pressure residuals to 10^{-6}	3560 [s] for 20000 time steps, to bring the average pressure residuals to 10^{-7}
CPU time/(iteration*grid cells)	0.781×10^{-5}	0.767×10^{-5}

COMPUTER RELATED AND GLOBAL QUANTITIES

The computer related quantities are reported in Table 1.

The computed torque for *case 1* is equal to 0.2427, while for *case 2* it is 0.2965. The efficiency measures the ratio of the mechanical energy produced over the hydraulic energy. We do not report it because inviscid simulations have been performed and in our case the computed efficiency would mainly provide an indication of the losses caused by the artificial dissipation introduced explicitely in the numerical procedure.

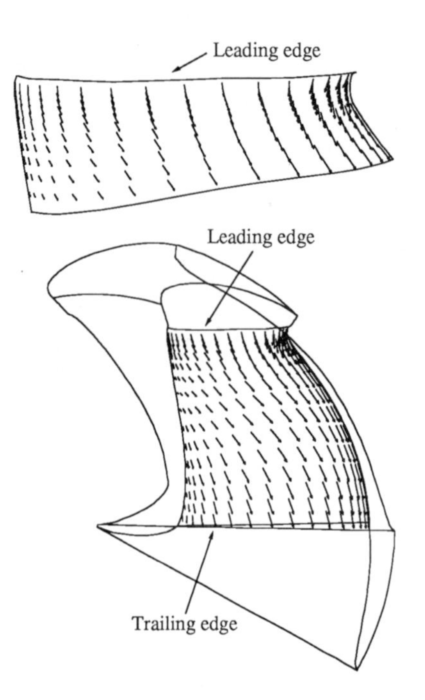

Fig. 7. Velocity fields on the blade suction side, case 1.

CONCLUSIONS

The quality of the results achieved by inviscid rotational simulation of the flow in Francis runner and distributor makes the technique particularly appealing to researchers and constructors alike. A weak point is constituted by the grid of the computational domain, over which a discrete solution is sought. The generation of what could be deemed as a "good" mesh has proven to be a very difficult task, involving several trial and error phases. Still the results obtained, for example on the distributor, were such that some errors on the leading edge of the blades were visible and propagated upstream. Although the global solution was only slightly affected this is a matter of concern, and a considerable amount of research time is being devoted to studying C and O meshes around turbomachinery blades [5] and the multiblock

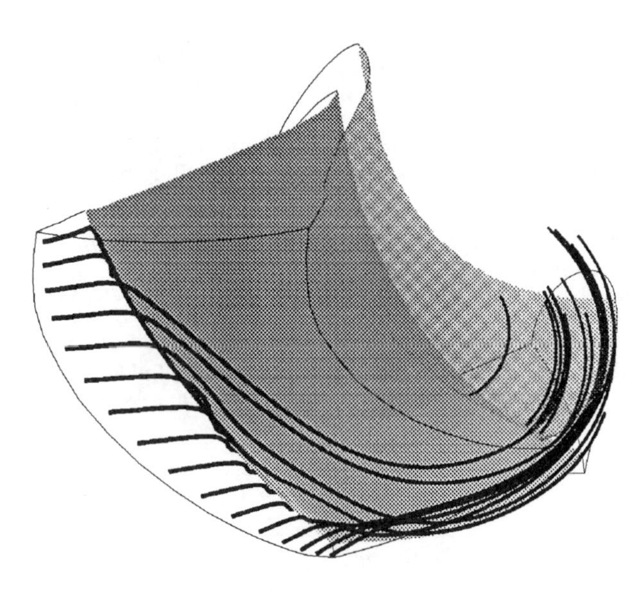

Fig. 8. Particle traces in the runner.

method, described by P. Eliasson in this Workshop.

The technique of artificial compressibility used within the finite volume formulation has demonstrated to be a robust and reliable method to achieve steady state incompressible flow solutions and it is likely to become a preferred tool for industrial applications.

ACKNOWLEDGMENTS

We wish to acknowledge Prof. I.L. Ryhming, director of IMHEF-Ecublens at EPFL, for providing both an ideal working environment and guidance. We are grateful for the help provided by our colleagues R. Richter, who developed the graphical visualization package CACTUS, and N. Ytterdahl, who wrote the program to compute particle traces. F. Thibaud assisted in the preparation of some 3D visualization pictures.

The ongoing research of IMHEF on hydraulic machines is supported by the *Ateliers de Constructions Mécaniques de Vevey S.A. (ACMV), Sulzer-Escher-Wyss Ltd. (SEW)* and the *Nationaler Energie-Forschungs-Fonds (NEFF).*

REFERENCES

[1] BOTTARO, A. : *"A note on open boundary conditions for elliptic flows"*, Num. Heat Transfer, Part B, 18 (1990), pp. 243-256.

[2] CHORIN, A.J. : *"A numerical method for solving incompressible viscous flow problems"*, J. Comp. Physics, 2 (1967), pp. 12-26.

[3] ELIASSON, P. : *"Navier-Stokes solutions for laminar incompressible flow over a NACA0012 airfoil and a backward facing step"*, FFA Report TN 1987-50, 1987.

[4] NEURY, C. : *"3-D mesh generation for calculating flow through radial-axial turbines"*, in Numerical Grid Generation in Computational Fluid Dynamics, J. Häuser and C. Taylor eds., Pineridge Press, 1986, pp. 387-398.

[5] NEURY, C., BOTTARO, A. : *"Influence of mesh type in the simulation of hydraulic machines"*, in Modern Technology in Hydraulic Energy Production, 15th IAHR Symposium, Belgrade, Yugoslavia, Sept. 1990, 1, paper C4.

[6] SAXER, A., FELICI, H. : *"Etude numérique d'écoulements internes, incompressibles et stationnaires par les équations d'Euler"*, IMHEF Report T-87-4, 1987.

[7] SAXER, A., FELICI, H., NEURY, C., RYHMING, I.L. : *"Euler flows in hydraulic turbines and ducts related to boundary conditions formulation"*, in Proc. Seventh GAMM Conf. Num. Meth. Fluid Mech., M. Deville ed., NNFM 20, Vieweg Verlag, 1988, pp. 343-354.

[8] THIBAUD, F., DROTZ, A., SOTTAS, G. : *"Validation of an Euler code for hydraulic turbines"*, AGARD Symp. Validation Comp. Fluid Dyn., AGARD-CP-437, 1 (1988), pp. 27-1 to 27-14.

Numerical Solution of the Incompressible Euler Equations in a Water Turbine Using a Multi-Block Approach

Peter Eliasson
FFA The Aeronautical Research Institute of Sweden
161 11 Bromma, Sweden

SUMMARY

The flow in the distributor and the runner of the 1989 GAMM workshop water turbine has been predicted by numerical solutions of the incompressible Euler equations. The Euler solver uses the artificial compressibility technique in order to find a steady solution. The governing equations are numerically solved using a finite volume discretisation in space and explicit Runge-Kutta integration in time. Artificial damping must be added to the numerical scheme.

A multi-block technique is used for the simulation of the flow in the distributor and the runner simultaneously. At the boundary between the distributor and the runner, averaging in the tangential direction is made in order to find a steady solution. The numerical results proved to be in good agreement with existing experiments and with other numerical data.

INTRODUCTION

The artificial compressibility technique, originally proposed by Chorin [2], has been applied to a number of different incompressible flow cases, e.g. pump calculations [3], calculations around cars [1], incompressible flow over delta wings [8], viscous two-dimensional flows [4] and even turbulent flows [5,6]. The method is found to be very robust for all of these different flow cases so it was a natural choice to apply it to the 1989 GAMM water turbine.

The incompressible Euler equations are numerically solved using a second order finite volume discretisation in space and a four stage explicit Runge-Kutta integrator in time. Since central differences are used and no physical dissipation is present, the scheme is damped by a fourth order artificial damping.

A multi-block technique is used to simulate the flow in the distributor and the runner simultaneously. Three blocks were used, one block for the stay vane, one block for the guide vane and one block for the runner (Figs. 1 and 2). Since the guide vane is fixed and the runner is rotating, averaging in the tangential direction is made at the boundary between the guide vane and the runner in order to find a steady solution. This means that there is no tangential variation of the flow variables at this boundary.

The numerical results proved to be in very good agreement with existing experiments for the distributor and on the inlet to the runner. The results for the runner were also good, with some minor discrepancies at the outlet which could probably be explained by the neglected tangential variation at inlet boundary of the runner.

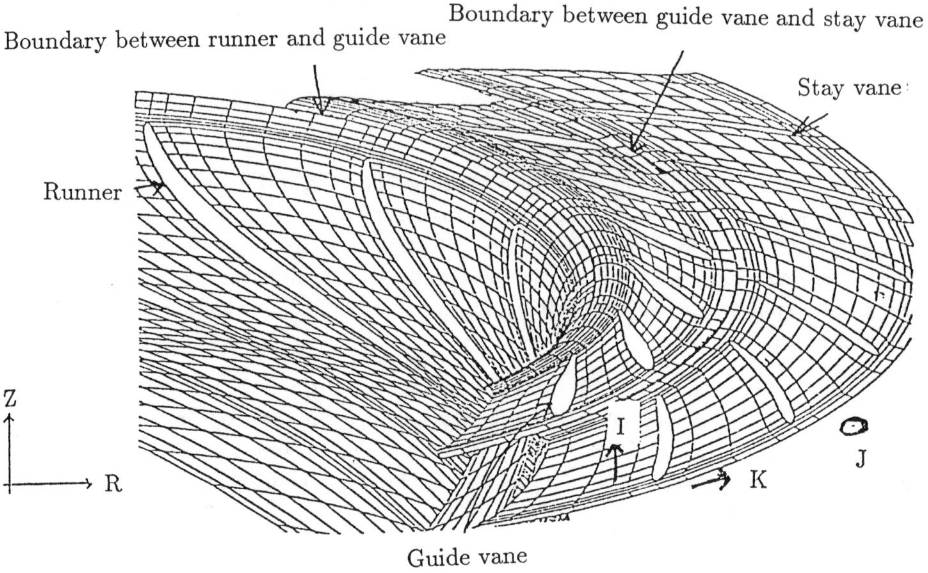

Boundary between guide vane and stay vane
Boundary between runner and guide vane
Stay vane
Runner
Z
R
I
J
K
Guide vane

Figure 1: Parts of the meshes used on the hub

MATHEMATICAL MODEL

In order to be able to solve the incompressible Euler equations by a time marching procedure, an artificial time dependent term is added to the continuity equation. This is called the artificial compressibility technique and was first suggested by Chorin[2]. The equations become :

$$\frac{1}{\rho_0}\frac{\partial p}{\partial t} + c^2 \nabla \cdot \vec{u} = 0 \quad ,$$

$$\frac{\partial \vec{u}}{\partial t} + (\vec{u} \cdot \nabla)\vec{u} + \frac{1}{\rho_0}\nabla p = -\vec{\omega} \times (\vec{\omega} \times \vec{r}) - 2(\vec{\omega} \times \vec{u}) \tag{1}$$

where ρ_0 is the constant density, $\vec{u}$ is the velocity vector, p is the pressure and c is a parameter chosen to optimize convergence. $\vec{\omega}$ is the constant angular velocity vector

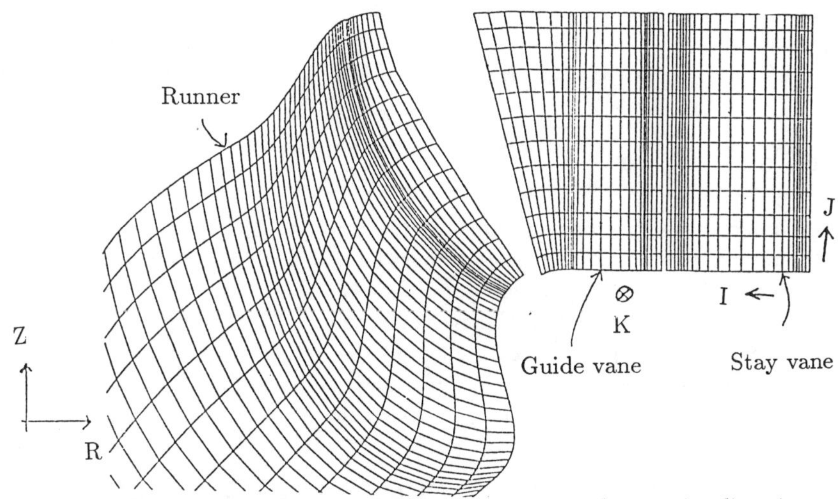

Figure 2: Parts of the meshes in streamwise and spanwise directions

because the calculations are made in a rotating coordinate system. As the steady state is obtained, it is readily verified that the steady Euler equations are solved. The optimal value of c^2 was analyzed by Rizzi & Eriksson [8] and was found to be approximately equal to the square of the freestream velocity.

Introducing a cartesian coordinate system and applying the integral formulation of (1), the equations can be written

$$\int_\Omega \frac{\partial \vec{q}}{\partial t} dV + \oint_{\partial\Omega} \mathbf{H}(\vec{q}) \cdot d\vec{S} = \int_\Omega \vec{b} dV \quad , \tag{2}$$

$$\vec{q} = \begin{pmatrix} p/\rho_0 \\ \vec{u} \end{pmatrix} \quad , \quad \mathbf{H}(\vec{q}) = \begin{pmatrix} c^2\vec{u} \\ \vec{u}\vec{u} + p/\rho_0\mathbf{I} \end{pmatrix} \; , \; \vec{b} = \begin{pmatrix} 0 \\ -\vec{\omega} \times (\vec{\omega} \times \vec{r}) - 2(\vec{\omega} \times \vec{u}) \end{pmatrix}$$

where $d\vec{S}$ is the outward pointing normal surface vector.

NUMERICAL METHOD

Spatial discretisation

A centered finite volume method is used to spatially discretise the governing equations expressed in integral form (2). No details about the finite volume method will be presented here since the discretisation is straight forward and can be found in the literature, e.g. [1,2,3,4,8]. The finite volume method is second order accurate in space.

Time integration

The method to integrate ordinary differential equations, like the discretised version of (2) in space, is an explicit one step, four stage, first order accurate Runge-Kutta algorithm which allows a theoretical CFL limit of 3. It has earlier been shown that this algorithm is superior to the standard three stage, second order accurate Runge-Kutta scheme [4]. The convergence characteristics are improved by using local time stepping.

Artificial dissipation

The central differences in the finite volume discretisation of (2) give rise to oscillations which is the reason why numerical viscosity must be added to the scheme to damp the short wavelengths. The damping consists of differences which must be of higher order than the differences in the discretised numerical scheme. Thus, the total differencing operator $\vec{F}$ consists of the physical part $\vec{F}_{ph}$ and the numerical part $\vec{F}_n$. In interior cells the numerical damping is defined by a fourth order difference operator, and the semi-discrete approximation of the incompressible Euler equations can be written:

$$\frac{\partial \vec{q}_{i,j,k}}{\partial t} = \vec{F}_{ph}(\vec{q}_{i,j,k}) + \vec{F}_n(\vec{q}_{i,j,k}) \quad , \quad \vec{F}_n(\vec{q}_{i,j,k}) = -\Gamma(\delta_i{}^4 + \delta_j{}^4 + \delta_k{}^4)\vec{q}_{i,j,k} \qquad (3)$$

where $\delta_i \vec{q}_{i,j,k} = \vec{q}_{i+\frac{1}{2},j,k} - \vec{q}_{i-\frac{1}{2},j,k}$, and similarly for $\delta_j \vec{q}_{i,j,k}$ and $\delta_k \vec{q}_{i,j,k}$. $\Gamma = \epsilon_4 CFL/\Delta t$ with ϵ_4 a constant in the range 0.005 to 0.01. Δt is the local time step. Near boundaries, $\vec{F}_n$ is defined by non-centered differences [4,8] to ensure the dissipative property of damping.

Boundary conditions

The mesh used for the calculation of the distributor and runner for the 1989 GAMM workshop water turbine consists of three different blocks, one for the stay vane, one for the guide vane and one for the runner. Parts of the three mesh surfaces on the hub for a coarse mesh can be seen in Fig. 1. Because of symmetry, the calculations are only carried out in one block in the tangential direction. In Fig. 2 the meshes in the streamwise and spanwise direction can be seen.

At a solid wall there is no normal mass flux, $\vec{n} \cdot \vec{u}$. However, the pressure contributes to the momentum flux. The value at the wall is obtained by extrapolation from the interior. Experiments with the order of accuracy of the extrapolation procedure indicate very small variations. We have chosen a second order extrapolation.

On the inlet boundary, the inlet to the stay vane, the three velocity components were fixed according to experimental values on one vertical line and assumed to have no variation in the tangential direction. The pressure was extrapolated upstream. At the outlet of the runner, so called characteristic boundary conditions were used [3,8].

The boundary conditions between the blocks must be treated in such a way that a steady solution can be obtained. At the boundary between the stay vane block and the guide vane block, continuity is simply assumed. At the boundary between the guide vane block and the runner block (the guide vane is fixed in space, the runner is rotating), averaging of all flow variables $(p, \vec{u})$ in the tangential direction (K-direction) is done according to the following procedure (Fig. 3) :

$$u_{Rav} = \frac{1}{NK_R - 1} \sum_{K_R=1}^{NK_R-1} u_R(I_R = 1, K_R)\, cos\varphi_R(K_R) + v_R(I_R = 1, K_R)\, sin\varphi_R(K_R) \ ,$$

$$v_{Rav} = \frac{1}{NK_R - 1} \sum_{K_R=1}^{NK_R-1} -u_R(I_R = 1, K_R)\, sin\varphi_R(K_R) + v_R(I_R = 1, K_R)\, cos\varphi_R(K_R) \ ,$$

$$w_{Rav} = \frac{1}{NK_R - 1} \sum_{K_R=1}^{NK_R-1} w_R(I_R = 1, K_R) \ , \quad p_{Rav} = \frac{1}{NK_R - 1} \sum_{K_R=1}^{NK_R-1} p_R(I_R = 1, K_R)$$

where subscript R means runner. Corresponding formulas also hold for the guide vane when subscript R is substituted to G and $I_G = NI_G - 1$. $\varphi(K)$ is the tangential angle that varies in the K-direction. The values of the four flow variables in the dummy cells of each block can then be obtained :

$$\begin{aligned}
u_G(I_G = NI_G, K_G) &= u_{Rav}\, cos\varphi_G(K_G) - v_{Rav}\, sin\varphi_G(K_G) - \omega\, y_G(K_G) \ , \\
v_G(I_G = NI_G, K_G) &= u_{Rav}\, sin\varphi_G(K_G) + v_{Rav}\, cos\varphi_G(K_G) + \omega\, x_G(K_G) \ , \\
p_G(I_G = NI_G, K_G) &= p_{Rav} \ , \quad w_G(I = NI_G, K_G) = w_{Rav} \ , \\
u_R(I_R = 0, K_R) &= u_{Gav}\, cos\varphi_R(K_R) - v_{Gav}\, sin\varphi_R(K_R) + \omega\, y_R(K_R) \ , \\
v_R(I_R = 0, K_R) &= u_{Gav}\, sin\varphi_R(K_R) + v_{Gav}\, cos\varphi_R(K_R) - \omega\, x_R(K_R) \ , \\
p_R(I_R = 0, K_R) &= p_{Gav} \ , \quad w_R(I_R = 0, K_R) = w_{Gav} \ .
\end{aligned}$$

Note that the calculations for the runner are done in a rotating coordinate system (and that the axis of rotation is the z-axis), which is why ω appears in the formula above. In the spanwise direction (J-direction) the mesh lines on the two blocks coincide at the interface (Fig. 3).

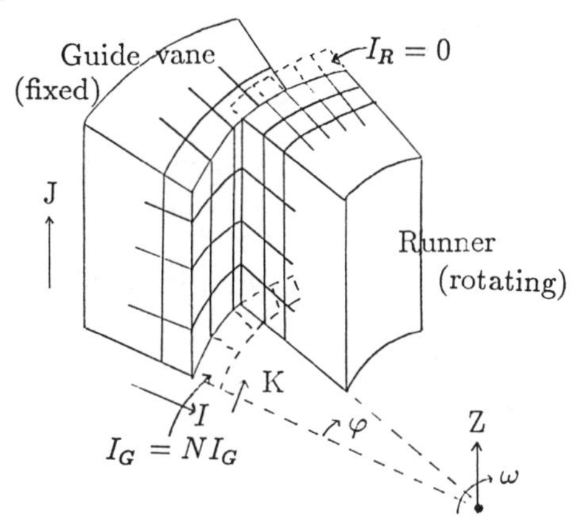

Figure 3: The boundary between the guide vane and the runner

The averaging procedure of the flow variables results in no tangential variation of the flow field on the boundary between the guide vane block and the runner block.

MESH GENERATION

All meshes used have been created with the same mesh generator. The mesh generator creates a number of locally two dimensional meshes with the arclength in the streamwise direction and the radius times the angle in the tangential direction as dependent variables. The $2D$-meshes are then transformed to the cartesian coordinate system and stacked on top of each other. The principles are very similar to those of Eriksson [7].

The blades in the stay vane and guide vane were cut off at the trailing edge. In order for the mesh generator to handle this, one extra point was added to the geometry outside the trailing edge (Fig. 4). This makes the trailing edge round when splines are used to interpolate the trailing edge points.

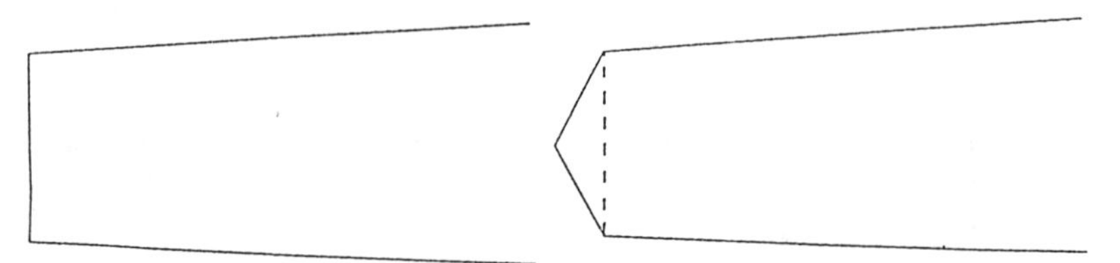

Figure 4: Modification of the trailing edge of the stay and the guide vane

RESULTS

The value of the parameter c^2 was chosen to $c^2 = 200$ according to given velocity profiles. The damping factor ϵ_4 used was $\epsilon_4 = 0.01$. The following computer related quantities were obtained:

$$Computer\,type : CRAY - XMP$$
$$No.\,of\,grid\,points : 2 \times (49 \times 25 \times 13) + 113 \times 25 \times 21 = 91175$$
$$CPU - time : 3178s,\ 2000steps$$
$$CPU - time/(No.\,of\,grid\,cells) : 3.90 \times 10^{-2}$$
$$CPU - time/(Iteration * No.\,of\,grid\,cells) : 1.95 \times 10^{-5}$$
$$CPU(scalar)/CPU(vector) = 132.87/37.27 = 3.56$$
$$No\,parallelisation\,speedup$$

The following global quantities for the runner were obtained:

$$Resultant\,nondimensional\,torque\ :\ 0.267$$
$$Efficiency\ :\ 0.944$$

The deviation from 1.0 for the efficiency is mainly due to the presence of artificial dissipation. The convergence diagram for the pressure can be seen in Fig. 5.

The agreement with existing experimental data for the distributor at inlet and outlet is very good. Looking at the pressure distributions for the stay vane and guide vane it is seen that the curves of the suction side and pressure side intersect close to the trailing edge. This can be explained by the round trailing edge. Other groups of investigators

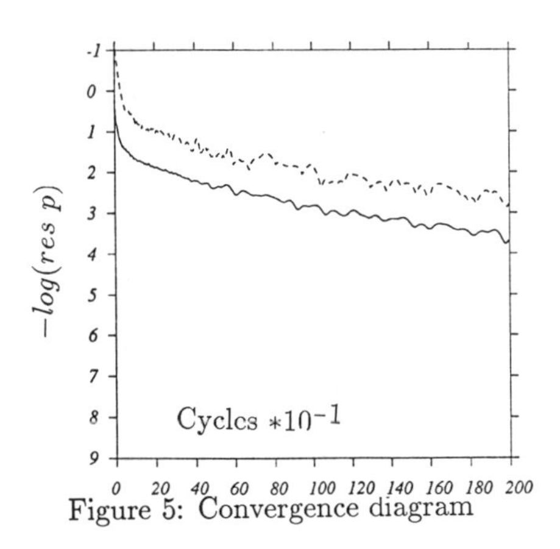

Figure 5: Convergence diagram

who applied a sharp trailing edge did not obtain this intersection. Questions may be raised whether an Euler calculation should be applied to a geometry having a round trailing edge where viscous effects are dominating. Is the Kutta-condition satisfied, is the flow oscillating ? There are no clear answers to these questions since it is not clear how the artificial dissipation acts for such a geometry. It is clear, however, that in this particular case, no problems with convergence were obtained and the agreement with experiments at the outlet of the distributor is very good.

For the runner the agreement with experiments is also good. The small discrepancies at the two measurement axis downstream the blade can probably be explained by the neglected variation in the tangential direction due to the averaging technique. This seems like a reasonable explanation since the required quantities, which were averaged in the tangential direction at the inlet of the runner, are very close to the experiments. It is important to remember that the experimental values on the axis that were supplied to the participants were also averaged values. Other groups that had no tangential variation of the flow variables on the inlet of the runner, also showed minor discrepancies downstream the blade. The exception was the Norwegian group that used characteristic boundary conditions at the inlet of the runner which allows a tangential variation. Their results downstream of the runner were very close to the experimental values which confirms that the tangential variation is indeed important.

The mesh size between the different groups varied a lot. Some of the groups used very coarse meshes, which resulted in poor agreement with the experiments. The results obtained in this paper were obtained in three steps. First a converged solution was obtained on a coarse grid, obtained by using every fourth point of the finest mesh in all three coordinate directions. The coarse grid solution was then interpolated to a medium mesh (every second point) and used as an initial condition to obtain the medium mesh solution. The converged solution on this medium mesh was then used as an initial solution for the finest mesh. The differences between the solutions on the two finer grids were very small which indicates that mesh independence was obtained. Large differences can, however, be seen between the solution on the coarsest mesh and the solutions on the finer grids. The explanation of the varying results between the different groups must then be a combination of the effects from the mesh and the boundary conditions.

The deviations between the computed and experimental pressure curves along suction and pressure lines on the blades could probably be explained by the presence of a boundary layer. The computed pressure curves were very similar to the curves of other groups using the same artificial compressibility technique.

CONCLUSIONS

The flow in the distributor and the runner of the 1989 GAMM workshop water turbine was predicted by numerical simulations using the incompressible Euler equations. The artificial compressibility technique was used.

A multi-block technique was employed, where the averaging in the tangential direction of the variables gave very good agreement with experiments and other numerical results as far as the distributor is concerned. The neglected variation of the variables in the tangential direction due to averaging, was found to give small discrepances with experiments at the outlet of the runner. The code could probably be further improved by looking into other ways of making the averaging. It is also very likely that O-meshes around the blades would improve the resolution with fewer mesh points.

REFERENCES

[1] BERGLIND, T.: "Incompressible Euler solution on a multi-block grid around a car-like configuration", FFA TN 1987-65, 1988.

[2] CHORIN, A.J.: "A numerical method for solving incompressible viscous flow problems", J. Comp. Phys. 2 (1967), pp 12-26.

[3] ELIASSON, P. and KROUTHÉN, B.: "The construction of an incompressible inviscid Euler solver for rotating systems and a comparative study of two Euler pump flow solutions", FFA TN 1987-03, 1987.

[4] ELIASSON, P.: "Navier-Stokes solutions for laminar flow over a NACA 0012 profile and a backward facing step", FFA TN 1987-50, 1987.

[5] ELIASSON, P.: "Solutions to the Navier-Stokes equations using a $k-\epsilon$ turbulence model", FFA TN 1988-19, 1988.

[6] ELIASSON, P., RIZZI, A. and ANDERSSON H.: "Time-marching method to solve steady incompressible Navier-Stokes equations for laminar and turbulent flow", Notes on Numerical Fluid Dynamics 24 (1989), pp 105-112.

[7] ERIKSSON L.E.: "Simulation of transonic flow in radial compressors", Computer Methods in Applied Mechanics and Engineering 64 (1987), pp 95-111.

[8] RIZZI, A. and ERIKSSON, L-E.: "Computation of inviscid incompressible flow with rotation", J. Fluid. Mech. 153 (1985), pp 275-312.

A STACKING TECHNIQUE FOR MULTISTAGE 3D FLOW COMPUTATION IN HYDRAULIC TURBOMACHINERY

E. Goede
SULZER ESCHER WYSS, Hydraulics Division
P. O. Box, CH-8023 Zurich

SUMMARY

The flow through the distributor and the runner of a Francis turbine has been calculated by use of a 3D Euler code. Within each part of the machine, i.e. the stay vane ring, the wicket gate as well as the runner, the flow has been calculated separately. To put these solutions together, a Quasi-3D stacking technique is applied. It is shown that this matching procedure is an appropriate way in order to avoid the use of empirical data in numerical flow simulation.

INTRODUCTION

Within the past five years great progress has been made in the application of CFD (Computational Fluid Dynamics) to hydraulic turbomachinery [2]. A great variety of different computer codes using different theories is now in use, and the whole spectrum from potential flow methods to Euler codes and even Navier Stokes solvers is operational [3,8]. However, it is common practice to focus the flow calculation on single parts of the turbomachine, for instance the runner, to be able to apply complex 3D-codes with high accuracy in reasonable time. Therefore at inlet or outlet of the computational domain boundary conditions must be specified which are a priori not known and which often are specified based on empirical data.

The purpose of this paper is to make a step towards the flow simulation for a complete hydraulic machine in order to avoid the need for empirical data in the application of CFD. In this context it is evident that, when determining the performance of a Francis runner, the distributor has to be included in the flow calculation, fig. 1. The advantage of doing this is that the boundary conditions at inlet of the stay vane ring can easily be derived from the global data corresponding to head and flow rate of the water. The flow field can be assumed to be constant over the cross section at stay vane inlet despite the variation of the flow velocities in the circumferential direction according to the measurements. This peripheral variation of the flow field depends on the geometry of the spiral casing and leads to unsteady effects at the runner. However, these effects are not subject to this paper.

***Fig. 1**: Geometry of distributor and turbine runner*

Another point is worth to be considered. For industrial applications it is important to work as economically as possible. This means for CFD to find the most effective theoretical method to solve the flow problem. In terms of economics such a method has to be as accurate as necessary and as simple as possible. The use of a complex method does not mean a priori that the best solution will be achieved. Instead, the theoretical method should be chosen such as to cover the main physical effects involved. Therefore, in the presented flow simulation method, a complex flow calculation theory (3D Euler) is combined with a relatively simple quasi-3-dimensional procedure to match the single flow fields to a complete solution.

To summarize, the main intention in performing this work was to simulate the flow from the beginning of the stay vane ring to the outlet of the turbine runner without using any empirical data serving for boundary conditions. The idea was to test the chosen method instead of tuning it with the provided experimental data. Whenever theoretical methods are used for design purposes, the validation of theory plays an important role [1] and is essential to achieve reliable tools for the layout of machine parts [4,7].

THEORETICAL BACKGROUND

The method used for simulation of the flow through the turbomachine is based on the Euler formulation of the equations of motion. This is desirable, since the flow inside the Francis turbine is highly vortical giving rise to flow phenomena that cannot be captured with potential methods. The method has been used with success for long time in aerodynamics [5]. Hence, it was possible to make use of experience made in airplane design aerodynamics and to use already successfully working algorithms while developing this method [6]. However, the extension of the computational method to incompressible internal flows in turbomachines required special measures, mainly the introduction of artificial compressibility and periodicity conditions.

The physical basis of the calculation procedure is the conservation of mass and momentum inside the computational domain, leading to the *Euler equations of motion*:

$$\int_{\partial\Omega} \mathbf{w.n}\, dS = 0$$

$$\frac{\partial}{\partial t}\int_{\Omega} \mathbf{w}\, dvol + \int_{\partial\Omega} [\mathbf{w}(\mathbf{w.n}) + (p/\rho)\mathbf{n}]\, dS = \int_{\Omega} \mathbf{f}\, dvol \quad .$$

f stands for the mass force vector including centrifugal, Coriolis and gravitational terms, if the governing equations are written for the relative system:

$$\mathbf{f} = -\omega \times \omega \times \mathbf{r} - 2\omega \times \mathbf{w} - \mathbf{g} \quad .$$

To solve the Euler equations a *time marching procedure* is applied. However, for incompressible flow, no time derivative appears in the continuity equation or, in other words, the continuity equation is not hyperbolic in time leading to a system of equations that is poorly conditioned for numerical solution. However, by adding a time-dependent pressure term (artificial compressibility) the hyperbolic character can be achieved artificially for the continuity equation. Now the system of equations can be written as follows [7]:

$$\frac{\partial}{\partial t}\int_{\Omega} \mathbf{q}\, dvol + \int_{\partial\Omega} \mathbf{H(q).n}\, dS = \int_{\Omega} \mathbf{F}\, dvol$$

with

$$\mathbf{q} = \begin{bmatrix} p/\rho \\ u \\ v \\ w \end{bmatrix}, \qquad \mathbf{H(q).n} = \begin{bmatrix} c^2\mathbf{w.n} \\ u\mathbf{w.n} + (p/\rho)\mathbf{n.e}_x \\ v\mathbf{w.n} + (p/\rho)\mathbf{n.e}_y \\ w\mathbf{w.n} + (p/\rho)\mathbf{n.e}_z \end{bmatrix}, \qquad \mathbf{F} = \begin{bmatrix} 0 \\ \mathbf{f.e}_x \\ \mathbf{f.e}_y \\ \mathbf{f.e}_z \end{bmatrix}$$

and

$$\mathbf{w} = u\mathbf{e}_x + v\mathbf{e}_y + w\mathbf{e}_z \ .$$

This system of equations can now be solved effectively by use of a time marching procedure. After a sufficient number of time steps the artificially introduced compressibility term vanishes, and the steady state solution for the pressure and flow field is reached.

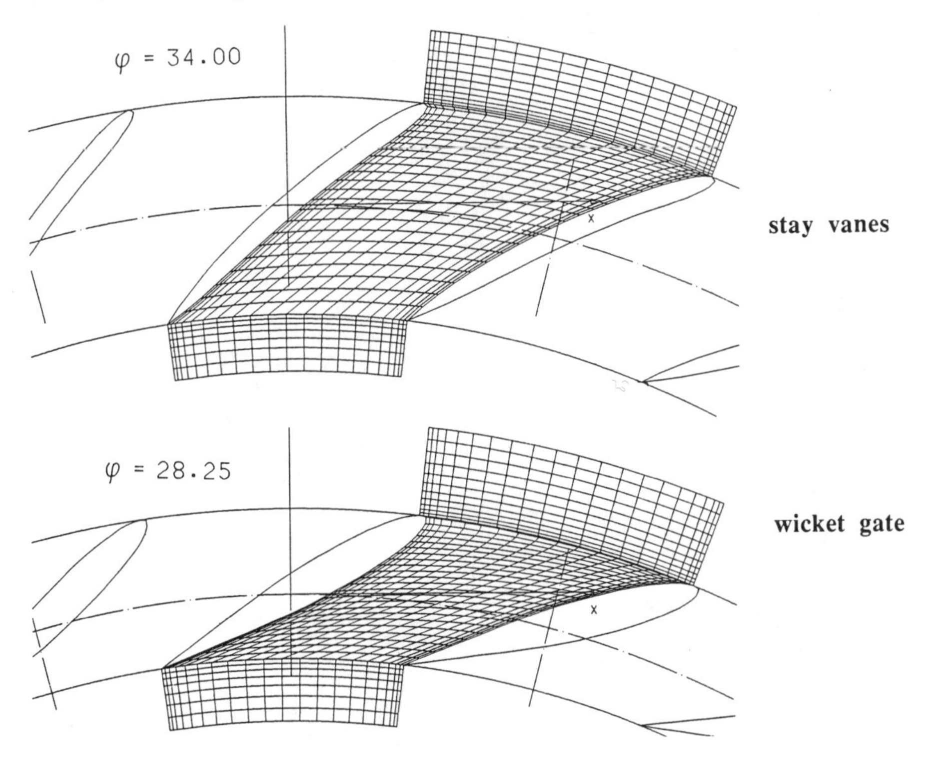

Fig. 2: *Computational mesh within stay vane and wicket gate blade cascades*

For discretization of the computational domain, necessary for numerical solution of the Euler equations, a *finite volume technique* is used, based on the integral formulation of the governing equations. The meshes used for the two blade cascades of the distributor are shown in fig. 2. The mesh size for both cascades is 42x20x12 elements corresponding to the orientation radial x peripheral x channel height, resulting in 10080 finite volumes. For the runner the discretized domain is presented in fig. 3 indicating the same mesh size but a difference in orientation: 42x12x20. Because of much more 3D effects to be expected for the turbine runner more elements are required from band to crown compared with the stationary cascades of the distributor.

All meshes are of the type H-H, which means that a H-mesh is used in the space between the blades as well as within the meridional flow channel. The size of the elements is optimized in such a way that, when getting close to the boundaries, the volume is reduced in order to be able to take into account high gradients within the flow field.

It is obvious that the flow calculation is carried out only for one blade passage, assuming that all other passages will have the same solution. The *boundary conditions* to be specified for both the stationary and the rotating parts are

- tangential flow on blade surface as well as on band and crown,
- periodicity conditions upstream and downstream of the blades,
- inlet flow and base pressure corresponding to discharge, head and rotational speed equivalent to the turbine operating point and to the installation level of the machine.

It is clear that, to run the above described method on a computer, a lot of data handling is needed. Consequently, the software for pre- and postprocessing is an inhouse solution, developed at *Sulzer Escher Wyss*, while the Euler-code was written at the *IMHEF/DME, Ecole Polytechnique Fédérale de Lausanne*.

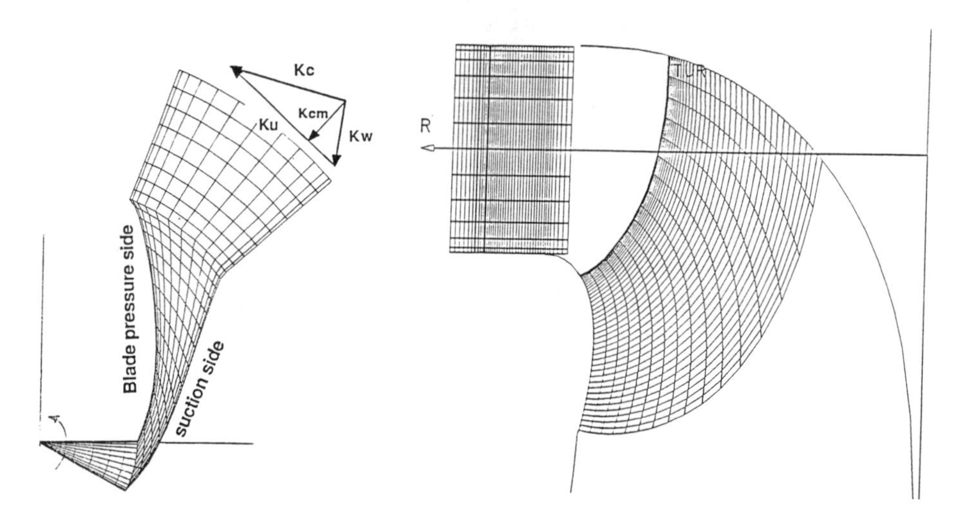

Fig. 3: *Runner inner blade mesh*

Fig. 4: *Wicket gate grid fitted into meridional channel*

STACKING OF SINGLE FLOW FIELDS

From the inlet of the distributor to the outlet of the turbine runner, the flow calculation has been carried out separately, in fact, within three parts of the machine:

- the stay vane ring,
- the (adjustable) wicket gate
- the runner.

To put all these single solutions together to a complete consistent flow field, the following steps have been applied:

1. The velocity distribution at each individual inlet is introduced downstream, beginning with a constant at stay vane inlet, using the result at stay vane outlet as input for the wicket gate, and so on. At stay vane inlet the boundary condition can be derived easily corresponding to the discharge and assuming zero incidence flow angle.

2. The pressure distribution (base pressure) at each individual outlet is introduced upstream, beginning with a constant at runner outlet equivalent to draft tube inlet, using the resulting distribution at runner inlet as input for wicket gate outlet, and so on. The base pressure at runner outlet is generally specified by the installation level of the turbine and the tailwater level. In the presented calculation the reference pressure is given by P_{ref}.

3 . If the trailing edge of the upstream part and the leading edge of the downstream part are located close together so that interference effects play an important role, an overlapping of the meshes to be linked is beneficial. The overlapping technique provides the possibility to specify the boundary conditions enough far away from the leading and trailing edges of the blades without influencing the flow field *locally* on the blading. For instance, the inlet condition for the (isolated) wicket gate can be specified upstream far away from the cascade inlet without any loss in accuracy, because the mean flow angle remains constant according to the flow in a logarithmic spiral. Consequently, the boundary conditions can be specified as constant in the peripheral direction and variable only across the meridional channel (Quasi 3D).

NUMERICAL RESULTS

In a turbine the pressure decreases from the entrance to the exit of the machine. This can be verified in fig. 5, in which the pressure distribution is given on each blade of the distributor and runner. The static pressure decreases from cp=1 at stay vane inlet to roughly cp=0 at runner outlet. The first considerable acceleration of the fluid takes place in the guide vane cascade, but the main acceleration is within the runner, where the power is generated. A 3-dimensional effect can be seen even in the guide vane, when comparing the pressure distribution at crown with that on the band. In addition, for the runner blade at band a low pressure peak can be seen close to the blade leading edge. This region is critical in terms of cavitation, in other words, if the pressure level is low enough and the local pressure drops below vapour pressure, cavitation will occur. This pressure level has been found to be equivalent to a Thoma factor of $\sigma = 0.14$. In a 3-dimensional representation (fig. 6) it can be recognized that the region with leading edge cavitation is found on blade suction side and on runner band as well.

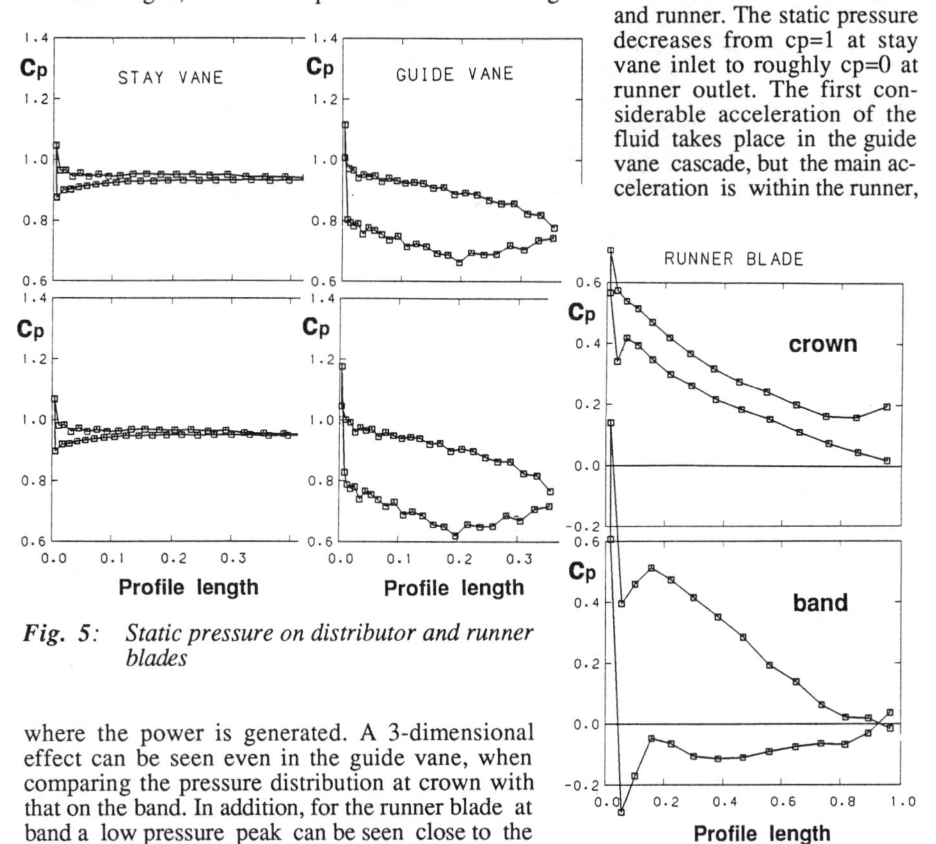

Fig. 5: Static pressure on distributor and runner blades

Essential for validation of the theory is the comparison of the calculated flow field with measurements upstream and downstream of the runner. Fig. 7 shows for the measuring planes *runner inlet* and *runner outlet* the computed velocity distributions across the meridional channel compared with the provided measurements (5-hole probe). To be sure, the theoretical curves represent mean values of the 3D results, which are impulse averaged along the peripheral direction.

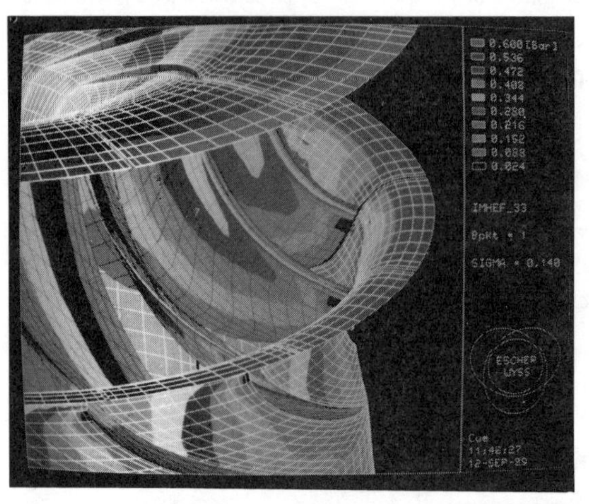

Fig. 6: Pressure field on runner surface for specified σ

The high accuracy of the flow computation is obvious, most important at runner outlet, the critical part of the validation. If downstream of the runner the numerical results are correct, there is evidence that all rotational effects inside the rotating part of the turbomachine are taken correctly into account, bearing in mind that between runner inlet and outlet the calculation is carried out in the relative system. In addition, if the difference in momentum between inlet and outlet of the runner is calculated correctly, the integration of the pressure distribution on the runner blading should lead to a realistic torque. For the runner torque and the corresponding hydraulic efficiency the following values have been found:

torque = 395 Nm (measurement: 375.5 Nm)

η_{hyd} = 95.2 %

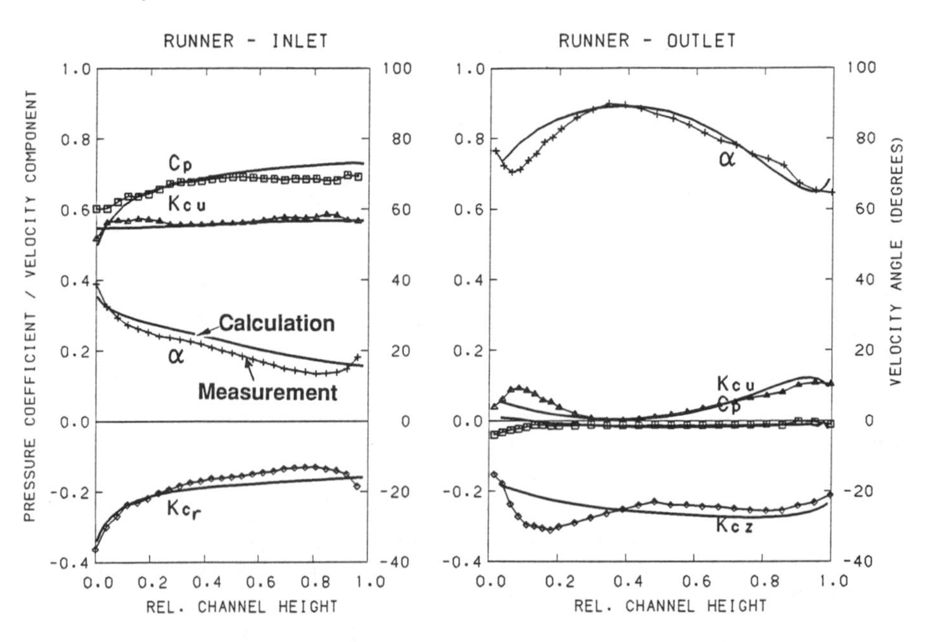

Fig. 7: Theory vs. experiment upstream and downstream of the runner

Finally, to show some details, the calculated flow field is presented in fig. 8 for inlet and outlet of the blade cascades for both wicket gate and turbine runner. The three columns stand for the flow along band, at mid span and along crown, each with four cross sections. From cross section 1 to 2 the acceleration of the water within the wicket gate can be seen according to the pressure drop shown in fig. 5. For sections 3 and 4 the velocity triangles are plotted, so that the comparison can be made between outlet of wicket gate (2) and inlet of the runner (3) in terms of the absolute flow as well as between cross sections 3 and 4 within the runner for the relative flow.

The results in all four cross sections seem to be consistent, and the turning of the absolute flow between outlet of wicket gate and outlet of the runner from high swirl to roughly zero swirl corresponds well with the turbine operating point of best efficiency. It is interesting to realize the great variation within the flow field upstream of the runner (cross section 3). At band the relative flow turns more than 20^0 between the blades. This is the reason for using

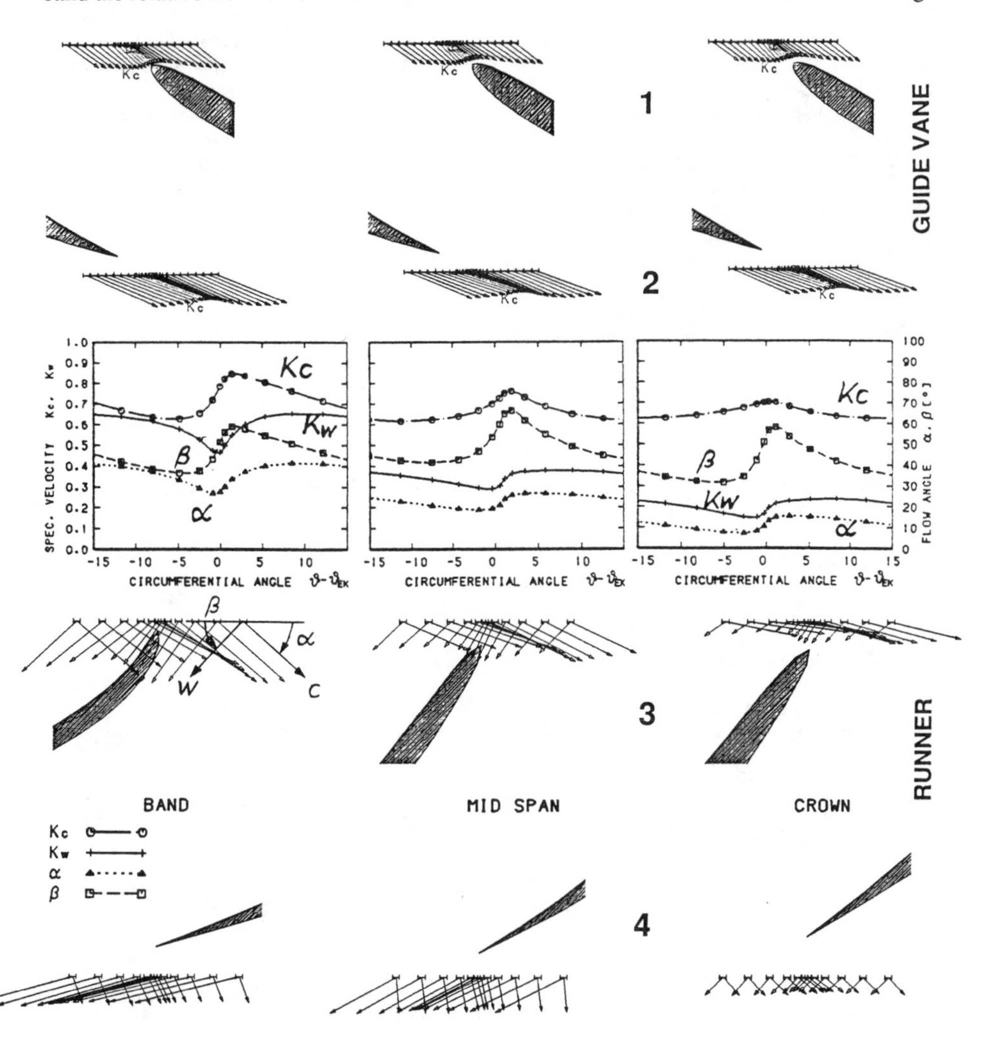

Fig. 8: Velocity triangles at leading and trailing edges of wicket gate and runner, for band, mid span and crown

an *overlapping* in the stacking technique of the computational grids. When getting with the inlet boundary of the mesh too close to the runner blade, the boundary condition, specified as constant in the peripheral direction, would influence the peripheral distribution of the solution. In fact, the solution would be flattened, peaks would be reduced, giving rise to a reduced accuracy of the flow simulation.

NOMENCLATURE

D_1	outside runner diameter	α	absolute flow angle
g	gravitational acceleration	β	relative flow angle
H	head of water	h_B	barometric head
Q	discharge	h_{Va}	vapour pressure head
T	torque	H_z	turbine level
ω	angular velocity	$NPSH = h_B - h_{Va} + H_z$	net positive suction head
$h = \omega T/\rho gHQ$	efficiency	$\sigma = NPSH/H$	Thoma's cavitation factor
$c_p = (P-P_{ref})/\rho gH$	pressure coefficient	**n**	outward unit vector normal to surface $\partial\Omega$
n	rotational speed		
$u_1 = \pi n D_1/60$	circumferential velocity	**f**	mass forces vector
c	absolute velocity	Ω	volume
$Kc = c/(2gH)^{1/2}$	specific absolute velocity	$\partial\Omega$	surface of the volume Ω
Kw	specific relative velocity		
Ku	specific circumferential velocity		
Kc_m	specific meridional flow velocity		
Kc_u	specific circumferential component of absolute velocity		

REFERENCES

[1] BRADLEY, R. G.: "CFD Validation Philosophy", AGARD Symposium, Lisbon, 1988, pp. 1-1 to 1-6.

[2] FISHER, R. K.: "Computerized fluid analysis of hydraulic turbines & pumps", an overview of the state-of-the-art, IAHR Symposium 1988, Trondheim.

[3] GOEDE, E., CUÉNOD, R.: "Numerical flow simulations in Francis turbines", Water Power & Dam Construction, May 1989, pp. 17-21.

[4] GOEDE, E., CUÉNOD, R., BACHMANN, P.: "Theoretical and experimental investigation of the flow field around a Francis runner", IAHR Symposium 1988, Trondheim, pp. 503-514.

[5] RIZZI, A., ERIKSSON, L.-E.: "Computation of flow around wings based on the Euler equations", J. Fluid Mech. vol. 148 (1984).

[6] SAXER, A., FELICI, H., NEURY, C., RYHMING, I.L.: "Euler flows in hydraulic turbines and ducts related to boundary conditions formulation". GAMM Conf., Louvain - la - Neuve, 1987.

[7] THIBAUD, F., DROTZ, A., SOTTAS, G.: "Validation of an Euler code for hydraulic turbines", AGARD Symposium, Lisbon, 1988, pp. 27-1 to 27-14.

[8] VU, T. C., SHYY, W.: "Viscous flow analysis for hydraulic turbine draft tubes", IAHR Symposium 1988, Trondheim, pp. 915-926.

3D EULER COMPUTATION OF THE FLOW INSIDE THE GAMM FRANCIS RUNNER

I. Grimbert, A. Verry
ELECTRICITE DE FRANCE - Direction des Etudes et Recherches
6 quai Watier, 78400 Chatou, France

E.M. El Ghazzani
METRAFLU
64 chemin des Mouilles, 69134 Ecully, France

SUMMARY

In the context of the GAMM Workshop, the flow in the Francis turbine runner has been calculated at EDF with a 3D steady Euler Code implementing a Clebsch Hawthorne formulation (splitting the velocity into a potential and a rotational part) and using a finite element discretization.

This paper presents a description of the method and analyses the results obtained with 2 different inlet data sets (experimental velocities and corrected velocities corresponding to measured flowrate).

INTRODUCTION

As buyer and operator of a wide range of large turbomachines, EDF has been developing and applying for many years flow analysis methods. Until recently QUASI 3D methods were common practice for computing the flow inside the impellers of hydraulic turbines, steam turbines and pumps for nuclear power plants.

Today a new generation of numerical tools is available at EDF to predict more accurately 3D flow patterns. They are the following :

- EULER 3D, which is a cost-effective code usable when the viscous effects are localized near the walls.
- N3S, which is a fully elliptic 3D viscous code, already available for fixed geometries such as draft-tubes. In the near future this code will be extended to compute the flow in a rotating frame of reference.

Since an extensive experimental program on the runner was planned by EPFL, we focused on the computation of the flow inside this part of the FRANCIS turbine with the EULER code.

DESCRIPTION OF THE EULER CODE

FORMULATION

The Euler code which has been used to compute the GAMM turbine runner has been developed at EDF in collaboration with METRAFLU.

It solves the 3D steady Euler equations through a finite element discretization. The method is based on the Clebsch-Hawthorne formulation, which decomposes the velocity vector into a potential component and a rotational one [1]; it includes an original treatment of the transport equations in order to lower numerical diffusion and a relaxation scheme (pseudo time integration) to improve the stability.

CLEBSCH FORMULATION

The Clebsch formulation derives from a variational principle (Hamilton's theorem) :

$$\delta \int_{\Omega} L \, d\Omega = 0 ,$$

where the Lagrangian L is defined as the difference between kinetic and potential energies.

In order to insure the equivalence between the primitive variable form of the Euler equations and the Clebsch form, it is necessary to add two constraints via Lagrange multipliers (ϕ,β) to verify mass and energy conservation [3].

This leads to the following variational form for a relative coordinate system :

$$\delta \int_{\Omega} \left(\frac{1}{2} \rho \vec{V}^2 + P^* + \phi \left[\vec{\nabla}.(\rho \vec{W}) \right] + \beta \left[\vec{\nabla}.(\rho P^* \vec{W}) \right] \right) d\Omega = 0 ,$$

where : $\vec{V}$: absolute velocity vector,

$\vec{W}$: relative velocity vector $\vec{W}=\vec{V}-\vec{U}$,

$\vec{U}$: entrainment velocity vector,

P^*: total relative pressure $P^* = P_s + \frac{1}{2} \rho \, (W^2 - U^2)$,

P_s: static pressure,

ϕ,β : Lagrange multipliers.

After integrating by parts, independent variations of the functional with respect to δV_i, $\delta\phi$, $\delta\beta$ and δP^* lead to the following equations of motion :

$$\delta V_i : \quad \vec{V} = \vec{\nabla}\phi + P^* \vec{\nabla}\beta$$

$$\delta\phi : \quad \vec{\nabla}.(\rho \vec{W}) = 0 \qquad (1)$$

$$\delta\beta : \quad \vec{\nabla}.(\rho P^* \vec{W}) = 0 \qquad (2)$$

$$\delta P^* : \quad \vec{\nabla}.(\rho \beta \vec{W}) = 1 \qquad (3)$$

which is the Clebsch formulation of the Euler equations.

NUMERICAL TREATMENT

Equation (1) is elliptic and may be discretized centrally with the Galerkin method.

Equations (2) and (3) are first order transport equations and require specific treatment due to their hyperbolic nature. Therefore, a second order scheme has been implemented by reapplying the convective operator $W.\nabla$. This leads to :

$$\vec{W}.\vec{\nabla}\left[\vec{\nabla}.(\rho\, P^*\, \vec{W})\right] = 0$$

$$\vec{W}.\vec{\nabla}\left[\vec{\nabla}.(\rho\, \beta\, \vec{W}) - 1\right] = 0\,.$$

This method, which lowers numerical diffusion and improves the scheme stability [1], allows a centered discretization. It is equivalent to discretize the initial first order equations with the following weight functions $N_i = W.\nabla N_i$.

This procedure increases the influence of upstream nodes in the flow direction.

PSEUDO TIME DEPENDENT INTEGRATION SCHEME

The system of equations is non linear and must be solved iteratively. The linearization is realized using an under-relaxation scheme based on a pseudo-time dependent formulation. The principle of this method is to add to each equation an evolution term which takes into account time derivatives of Clebsch variables [1,2].

FRONTAL SOLVER

The matrix form produced by the finite element discretization is solved using a frontal technique based on a Gaussian elimination procedure. The originality of this method is that it solves each equation as soon as it is assembled and thus storage is reduced.

This solver takes advantage of the symmetrical nature of the elementary matrices and of their identity for transport equations.

BOUNDARY CONDITIONS

The flow is supposed to be steady in the relative reference frame and periodical (i.e. identical for each successive blade passage). Therefore the computational domain reduces to a single blade passage shown on figure 1.

Inlet-Outlet

- Flow rate distribution is imposed as a surface integral upstream. Downstream, the flowrate is readjusted after each iteration through a scaling by a constant correction factor in order to fulfill the continuity requirement.

- The variables P^* and β being purely convected by the flow must be specified on the inlet surface only :
  - P^* distribution is known (physical quantity)
  - β is equal to zero if the flow is irrotational at inlet; otherwise β is determined through the normal component of vorticity : $\xi.n$.

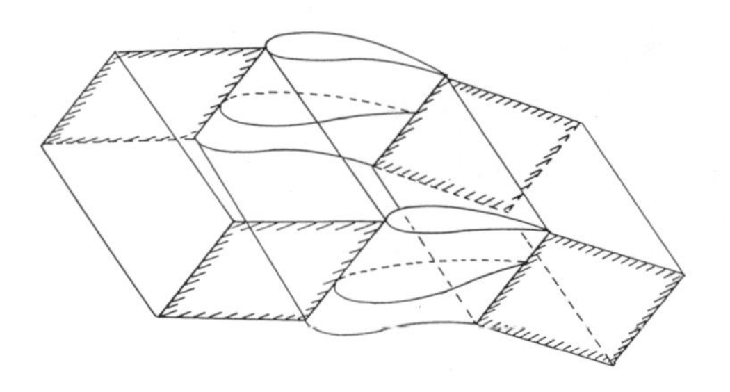

Fig. 1. Visualization of the computation domain

Periodicity

Requirement (on ϕ and β) on side surfaces : $\Delta\phi$ and $\Delta\beta$ (ϕ and β jumps) are determined from inlet conditions, upstream, and from Kutta-Joukowski condition at the trailing edge, downstream.

Solid boundaries

The flow is supposed to be tangent to the solid surfaces.

At one node of the domain (arbitrarily chosen)

The value of ϕ must be imposed at one node to insure the uniqueness of the solution and to prevent getting an ill conditioned matrix.

MESH GENERATION

The mesh is generated through a transfinite mesh generator. It is an H structured mesh with 20 noded quadratic elements (this Euler code may also deal with 8 noded linear elements

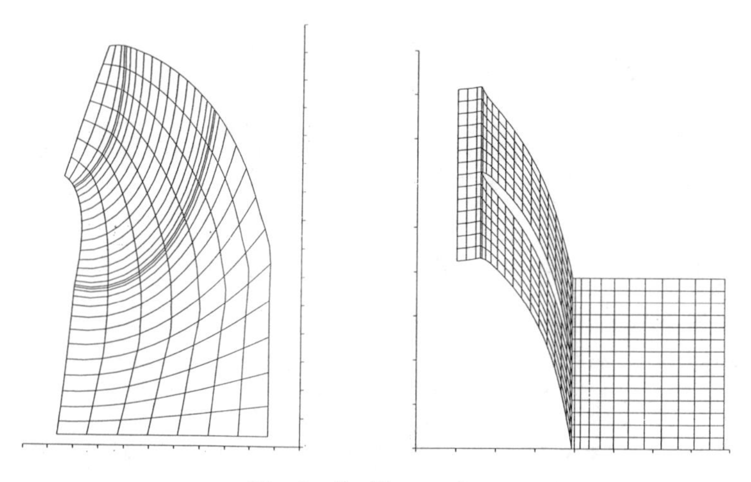

Fig. 2. Turbine mesh

or 27 noded quadratic elements).

Figure 2 presents a meridional and a blade to blade mesh view.

GAMM TURBINE RUNNER

COMPUTATION CHARACTERISTICS

For this application, the computation characteristics are the following :

Number of elements :	1862
Number of nodes :	9296
Computer type :	CRAY XMP (4-28)
CPU time/iteration :	48 s
CPU time/element :	0.4 s
CPU time for a complete computation :	12 mn

CONVERGENCE DIAGRAM

$$\varepsilon = \frac{Q_{inlet} - Q_{outlet}}{Q_{inlet}}$$

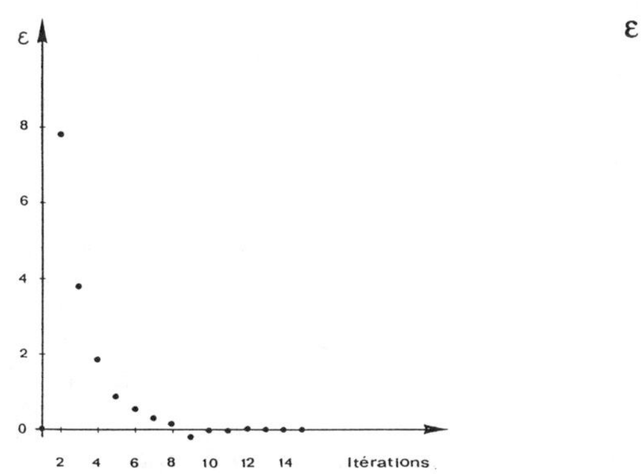

Fig. 3. Evolution of the mass flow error

At convergence, the flowrate conservation is insured on each transversal surface and the maximum variation is $\varepsilon = 10^{-3}$.

BOUNDARY CONDITIONS

The choice of boundary conditions have been the main problem of this test case : integration of experimental normal velocities at inlet and outlet did not lead to the measured flow rate. Therefore, a computation has first been run with an inlet data set corresponding to the experimental velocity field. Then, EPFL proposed to scale up the inlet velocity distribution in order to verify the measured flow rate. So, a second computation was performed with modified inlet velocities :

$$V'_{upstream} = 1.096\ V_{upstream}\ .$$

The inlet angles remained the same. Unfortunately, by doing so, the inlet kinetic momentum is no more consistent with the torque measurements.

COMMENTS ON THE RESULTS

- Both computations exhibit a high hydrodynamic load at the shroud due to a locally high incidence. This phenomenon is emphasized with the modified velocity distribution at inlet (figure 4). Such velocity distributions can lead to high losses in this area.
- Pressure distributions on the suction side of the blades show that cavitation should appear early at the inlet near the shroud, the pressure level being much lower there than anywhere else, especially with the modified velocity field (figure 4).
- Downstream the impeller, the computation with modified inlet velocities should only be considered. The comparison between experimental velocities and computed ones shows slight discrepancies on the axial velocity field. However, tangential velocities are well predicted. Therefore, the value of the hydraulic torque on the runner would have been accurately predicted if the inlet kinetic momentum had been the right one.

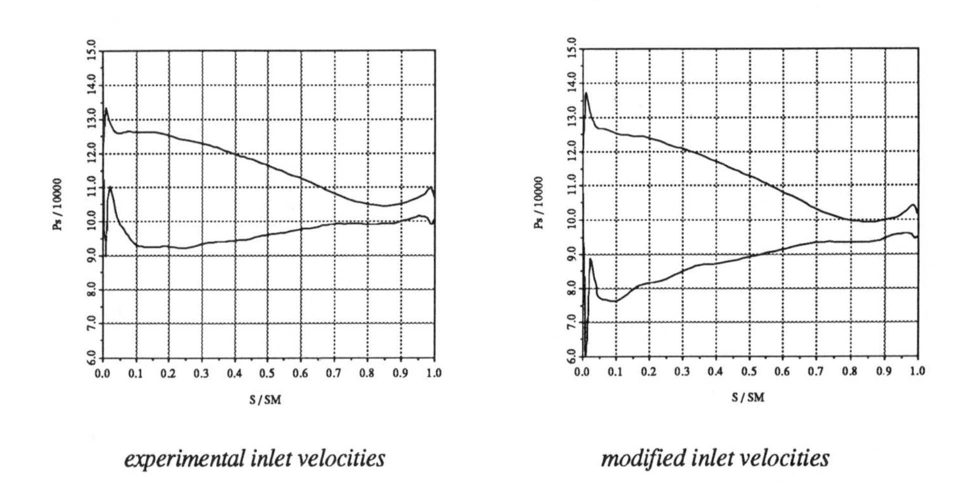

experimental inlet velocities *modified inlet velocities*

Fig. 4. Static pressure distribution on the blades near the shroud

CONCLUSION

This 3D Euler Code is now a cost effective and easy to implement industrial tool :

- The Clebsch-Hawthorne formulation lowers the order of the problem and the use of a frontal solver saves CPU time.
- The associated preprocessor quickly generates the mesh; the finite element discretization permits the treatment of complex geometries with any type of mesh (structured or unstructured).

It has already been validated on different machines (mixed flow pump, Stanitz duct.[2]).

The GAMM turbine will be an interesting and well documented test case when all the experimental results are available.

The GAMM workshop has been an opportunity to learn how to use in a better way the EULER codes. Particularly, it has been shown, once more, the strong influence of the upstream conditions for a runner computation and consequently, the interest of coupling it with a distributor calculation.

REFERENCES

[1] AKAY, H.U., ECER, A. : "Finite element solution of steady inviscid flows in turbomachinery", in Thermodynamics and fluid mechanics of turbomachinery, Ucer A.S., Stow P., Hirsch C. eds., Martinus Nyhott Publishers, 1985.

[2] CAUDIU, E., EL GHAZZANI, E.M., GRIMBERT, I., PHILIBERT, R., VERRY, A. : "3D flow computation in turbomachinery", I.A.H.R. Symposium Trondheim, 1988.

[3] SELIGER, R.L., WHITHAM, G.B. : "Variational principles in continuum mechanics", Proceedings of Royal Society A, London, 305 (1968), pp. 1-25.

3D N-S FEM FLOW ANALYSIS IN DRAFT TUBE

T.Kubota, K.Toshigami, R.Suzuki
Fuji Electric, Kawasaki
1-1 Tanabeshinden, Kawasaki-ku, 210 Kawasaki, Japan

SUMMARY

The three dimensional flow in the draft tube of the Francis turbine for the GAMM workshop '89 was analysed with an FEM Navier-Stokes code using constant eddy viscosity. A smaller value of the eddy viscosity was used in the near wall region in comparison to that used in the core region. Some illustrations are presented, which visualize the complex three dimensional flow field.

BASIC EQUATIONS AND MAIN FEATURES OF THE NUMERICAL METHOD

In our DRAFT TUBE analysis, the Navier-Stokes equations and the continuity equation for the three dimensional steady viscous incompressible flow are employed as the fundamental equations. The inclusion of turbulence models is limited in our approach, and hence, only an isotropic eddy viscosity coefficient is allowed, which, in addition, is assumed to be constant.

These equations are approximated by a partial upwind finite element method [4,5,6] using Bercovier-Pironneau elements [1,2] of 10-nodes-tetrahedral type; i.e. a conforming linear tetrahedral element is used to approximate the pressure, while over each sub-element of an element the velocity components are linearly interpolated.

For solving the non-linear algebraic equations derived from the above mentioned finite element approximation, a Newton-like method is adopted. In each iteration step of the method, Hood's wave front solver [3] is adopted to solve the linearized equations.

Considering these properties of the approach, the strategy for analyzing the GAMM DRAFT TUBE problem can be described as follows. By starting from a Stokes flow analysis, the eddy viscosity is gradually reduced from 10^{-2} to 3×10^{-4} (m^2/s) in the core region, while it is reduced from 10^{-4} to 10^{-5} (m^2/s) in the near wall region as shown in Table 2.

COMPUTATIONAL CONDITIONS

Figure 1 shows an overview of a finite element mesh using tetrahedral elements. In the circumferential direction, the region is equivalently divided into 12 parts. In the axial direction it was divided into 21 parts, and in the radial direction into 5 to 7 parts. Considering the existence of a strong vortex at the center of inlet cone, grids are concentrated at the neighborhood of the axis. In the bend and in the outlet cone, grids are spaced almost equally in the radial direction; besides, a fine mesh is used in the near wall region.

The level of the eddy viscosity is 1 or 2 orders of magnitude less in the near wall region than in the inner region. Therefore, two distinct values of the eddy viscosity are used.

The boundary conditions are summarized as follows.

(1) All velocity components on the wall are set to be zero.

(2) At the outlet, the pressure is set to be constant and a Neumann condition is applied for all velocity components.

(3) At the inlet, all velocity components are given. But the radial velocity component is slightly modified from the value specified by EPFL. Negative radial velocities near the inlet center are replaced with zero velocities, and the velocity near the wall is made tangent to the wall.

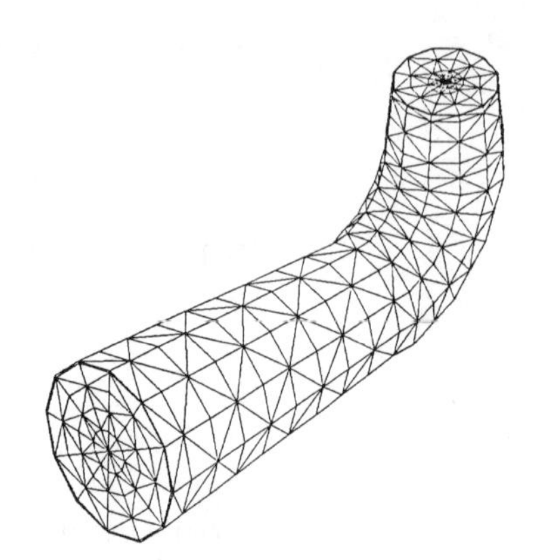

Fig. 1 Finite element mesh

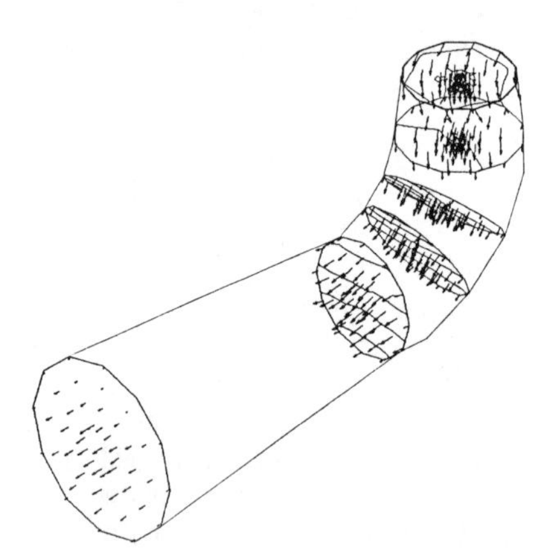

Fig. 2 Velocity and static pressure

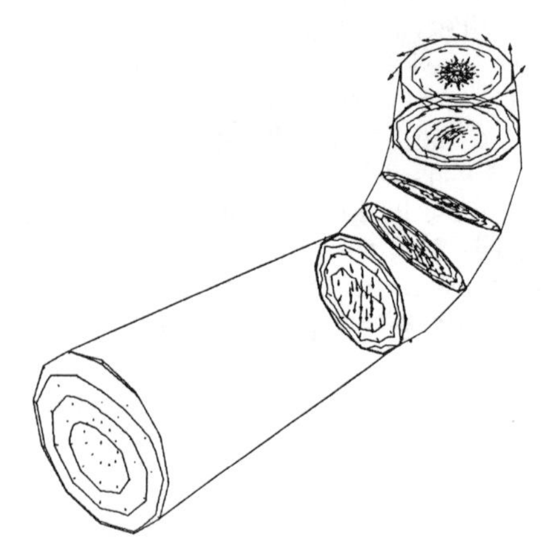

Fig. 3 Secondary flow and through flow

RESULTS

In order to visualize the complex three dimensional flow field in the draft tube, Figs. 2 and 3 are shown as an example. Figure 2 shows the velocity/pressure field. Arrows indicate velocity vectors and contours indicate constant static pressure. Figure 3 illustrates the secondary flow vector field and the contours of constant through flow velocity.

This calculation is carried out using a super computer with vector processors. Table 1 shows the computer related quantities and table 2 shows the convergence history.

Table 1 Computer related quantities

Computer type	FACOM VP 50 (OSIV/F4 MSP+VPCF)
No. of grids	9729
CPU time	16336 sec (*1)
(CPU time)/iteration	1257 sec (*2)
(CPU time)/(No.of grids)	1.68 sec
(CPU time)/(iteration∗No.of grids)	0.129 sec
Rate of vectorization	more than 8 times
Parallelization speedup	no parallelization

(*1) Including CPU time for Stokes flow analysis.
(*2) Stokes flow analysis was counted as the first iteration.

Table 2 Convergence history

Step	Order of the eddy viscosity (m^2/sec)		No. of iterations	Error (*3)	Supplement
	Near wall	the other			
—	10^{-4}	10^{-2}	——	———	Stokes flow
1	10^{-4}	10^{-2}	3	1.35×10^2	
2	10^{-4}	10^{-3}	3	1.07×10^2	
3	10^{-5}	10^{-3}	3	1.07×10^2	
4	10^{-5}	10^{-4}	3	1.07×10^2	

(*3) The error denotes $q / (1-q) * \| u^m - u^{m-1} \|$,

where $q = \max\left(\frac{\| u^m - u^{m-1} \|}{\| u^{m-1} - u^{m-2} \|}, \frac{\| u^{m-1} - u^{m-2} \|}{\| u^{m-2} - u^{m-3} \|} \right)$,

u^m is the vector of all degrees of freedom at the final iteration m and $\| \ \|$ denotes the Euclidian norm.

REFERENCES

[1] BERCOVIER, M., PIRONNEAU, O. : "Error estimates for finite element method solution of the Stokes problem in the primitive variables", Numer. Math. **33** (1979), pp. 211-224.

[2] GLOWINSKI, R., PIRONNEAU, O. : "On a mixed finite element approximation of the Stokes problem (I)", Numer. Math. **33**(1979), pp. 397-424.

[3] HOOD, P. : "Frontal solution program for unsymmetric matrices", Int. J. Num. Meth. Engng. **10**(1976), pp.379-399.

[4] IKEDA, T. : "Maximum principle in finite element models for convection-diffusion phenomena", Lecture Notes in Numerical and Applied Analysis, 4, Kinokuniya/North-Holland, 1983.

[5] KANAYAMA, H., TOSHIGAMI, K., MOTOYAMA, H. : "Three-dimensional air flow analyses in clean rooms by a finite element method", Theoretical and Applied Mechanics **36**(1988), pp. 35-46.

[6] KANAYAMA, H., TOSHIGAMI, K. : "Steady viscous incompressible flow analysis by a partial upwind finite element method", Computational Mechanics '86, pp. VII-163-168(1986).

COMPUTATION OF 3D FLOW FIELD IN A DRAFT TUBE

B. Lazzaro - P. Riva
RIVA HYDROART S.p.A.
Via Stendhal 34 - 20144 Milan, Italy

SUMMARY

The three-dimensional flow field inside the draft tube proposed by the GAMM WORKSHOP 1989 organizers is here simulated with FIDAP, a fluidodynamic program based on the finite element method. The mesh has been generated with the aid of FIDAP Meshing module. The solution method is based on the Navier-Stokes equations, taking into account turbulence with the k-ϵ model. The numerical analysis was carried out solving sequentially the submatrices that include only the unknowns of one conservation equation. Results show wide areas of secondary flow extended to the outlet section, which affect pressure recovery.

INTRODUCTION

The program utilized for the computation of the draft tube is FIDAP, developed by F.D.I.Inc.(USA). FIDAP is a commercial general purpose CFD code based on the FEM technique, dedicated to the study of several classes of incompressible fluid flows. The present paper may be also regarded as a test to validate a commercially available code in the hydraulic machinery field.

FLOW FIELD ANALYSIS

Basic Equations: For isothermal flows only mass conservation and Navier-Stokes form of the momentum equation are taken into account for the solution of the laminar flow field:

$$\nabla \cdot u = 0 \quad , \tag{1}$$

$$\rho\left(\frac{\partial u}{\partial t} + u \cdot \nabla u\right) = - \nabla p + \rho f + \rho g + \mu \nabla^2 u \quad . \tag{2}$$

The turbulent motion is modelled according to the two equations k-ϵ model [1], where k (turbulence kinetic energy) and ϵ (viscous dissipation of k) are described in the following forms:

$$k = 1/2\ \overline{u_i\ u_i}\quad , \tag{3}$$

$$\epsilon = \nu\ \overline{u_{i,j}\ u_{i,j}} = \nu\ \frac{1}{\Delta t} \int_t^{t+\Delta t} \hat{u}_{i,j}\ \hat{u}_{i,j}\ dt\quad . \tag{4}$$

Two transport equations can be derived for k and ϵ from the Navier-Stokes equations. Application of a number of modeling assumptions simplifies these two equations in the well known equations of the k-ϵ model, which contains empirical constants. It must be noted that an accurate calibration of these constants, according to specific experimental results, may lead to some improvements of the numerical solution.
In order to avoid excessive mesh density and consequent computation time, the universal law of the wall [1] is applied in the field included between the boundary of the viscous sublayer and the actual wall. It may be expressed as:

$$\frac{u_{res}}{u_\tau} = \frac{1}{\kappa} \ln\left(\frac{y u_\tau E}{\nu} \right)\quad . \tag{5}$$

Mesh Creation: The draft tube mesh has been entirely developed utilizing the capabilities of the FIDAP Meshing module.
The creation was done starting from a very simple planar topological section, to the nodes of which the XYZ coordinates of the corresponding points on the inlet section were associated. The grid was then extended in the topological space with other eleven sections to which the XYZ coordinates of the real sections nodes were linked (see fig. 1).

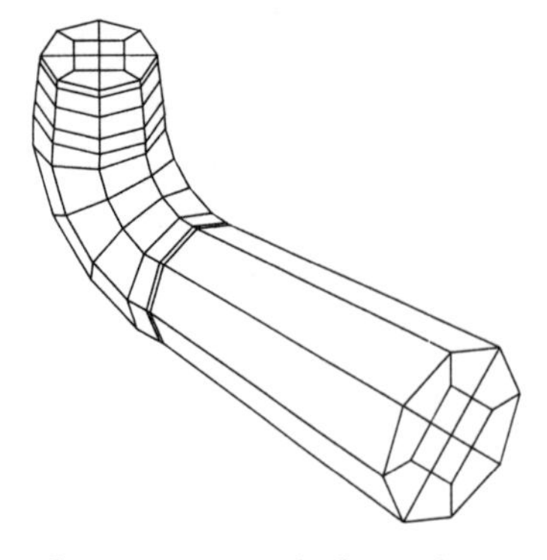

Fig. 1 - Unexpanded mesh

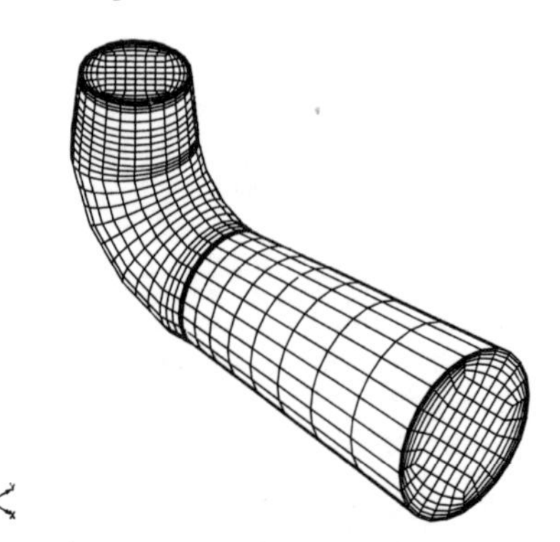

Fig. 2 - Final mesh

Starting from this mesh the nodal density has been strongly increased expanding the topology, without the need of addition of coordinates for the new nodes (see fig. 2).
Flow field discretization is done with 8-nodes brick elements, chosen in reason of the reduced total number of degrees of freedom and following reduced CPU time with respect to 27-nodes bricks.

Boundary Conditions: Experimental velocity profiles at the inlet section were introduced by means of subroutines linked with FIDAP's Main. Normal velocities at the draft tube wall have been assumed zero.
k and ϵ were inserted at the inlet in accordance with the following equations [1]:

$$k = (Lm \cdot du/dy)^2 / \sqrt{c} \quad , \tag{6}$$

$$\epsilon = c \cdot k^2 / (Lm^2 \cdot du/dy) \tag{7}$$

where c is an empirical constant equal to 0.09. The mixing length (Lm) distribution utilized for the computation is obtained from the Nikuradse's universal law of velocity in pipe flow [2]:

$$Lm = R(0.14-0.08(1-y/R)^2-0.06(1-y/R)^4) \quad . \tag{8}$$

Since the law of the wall is employed to obtain the solution near the wall, a distance y = 0.001 m is set for each node on the wall. This is the distance where the general computation is being "lifted" from the wall and the calculation switches to the law of the wall. k and ϵ at the wall have small non zero first-attempt values and have been recomputed during the execution.
No explicit boundary conditions were imposed at the outlet.

Solution Method: The solution algorithm utilized for solving the non linear system of matrix equation is one of the several available in FIDAP and is called segregated solver.
This algorithm is essentially based on an implicit approach. It avoids the direct formation of a global system matrix. Instead, this matrix is decomposed into smaller submatrices each governing the nodal unknowns associated with only one conservation equation. These smaller submatrices are then solved in a sequential manner using direct Gaussian elimination. The amount of memory necessary to store the submatrices is considerably less than that needed to store the global matrix.

RESULTS

A preliminary computation has been carried out assuming laminar flow, with an imposed viscosity $\nu = 0.005\ m^2/s$ (Re=217), in order to provide a realistic initial estimate of the velocity field. The turbulent computation has been performed at a $Re = 1.08 \cdot 10^6$, based on an inlet volume flow rate of $0.372\ m^3/s$ and an inlet draft tube diameter of 0.437 m.

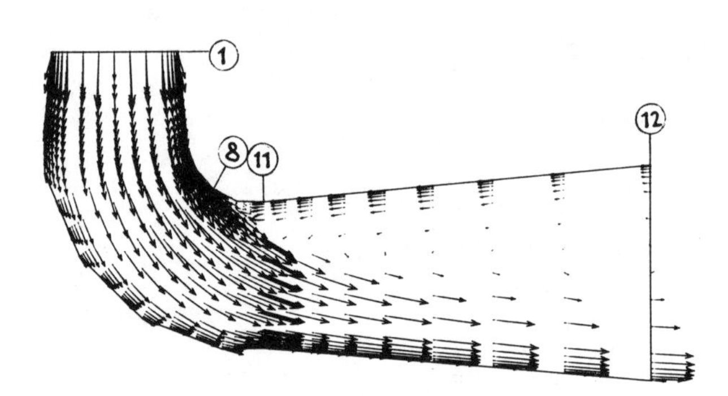

Fig. 3 - Velocity (elevation view)

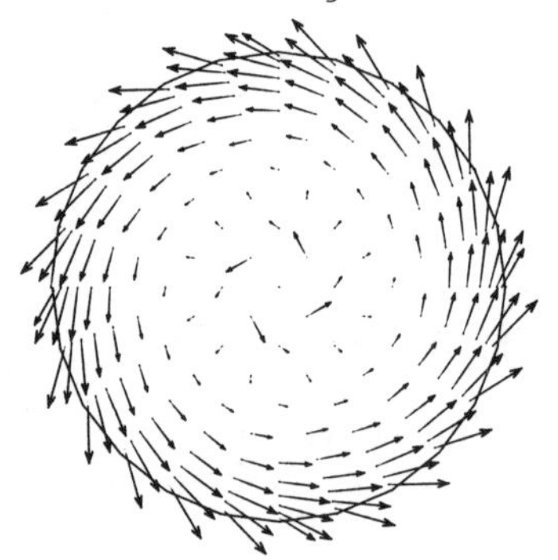

Fig. 4 - Velocity (section 1)

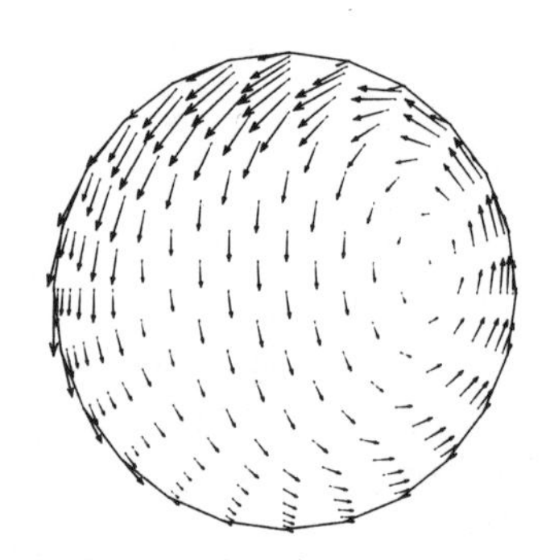

Fig. 5 - Velocity (section 8)

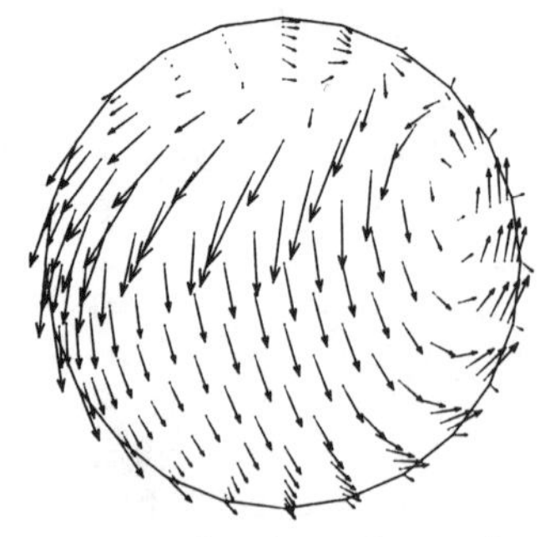

Fig. 6 - Velocity (section 11)

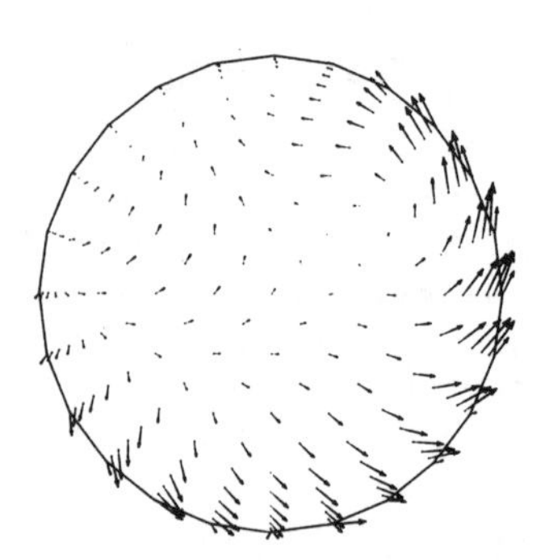

Fig. 7 - Velocity (section 12)

The results indicated a wide reversed flow region extended to the outlet section (see fig. 3-7, plotted in different scales). Such a behaviour affects the efficiency of the draft tube in terms of pressure recovery, as clearly visible in fig. 8; moreover it may be envisaged (even if not computed in the present calculation) that the extent of the recirculation is a function of Re, thus giving rise to an uneven behaviour of the draft tube for different flow-rates.

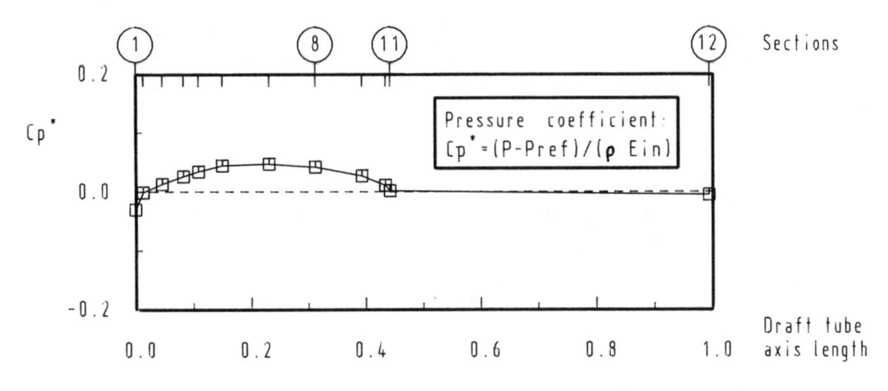

Fig. 8 - Pressure coefficient

As a consequence of the reversed flow, from the computational point of view, the boundary conditions of kinetic energy, dissipation and pressure cannot be specified with confidence. This may be one factor causing a rather slow convergence rate. The large size change of elements from the region described by section 10-11 to that in section 11-12 does not seem to cause any significant effect on the current accuracy of the velocity and pressure fields.
A deeper analysis of the results suggests a more accurate description of the central core region, in which a vortex breakdown may be present. A proper modelization of the vortex rope could also show an interaction between reversed flow and vortex rope, which may provoke an unstable behaviour of the draft tube itself and of the whole turbine.
The above considerations lead to uncertainty in result evaluation and validity of steady state computations whenever such a wide reverse flow region is detected.
For such reasons the validity of the computations performed has been checked only on a global magnitude as the discharge: volume flow rate at the inlet and outlet planes was computed to be within 3.4% and 5% of supplied values. The difference between inlet and outlet flow rates was within 2%. Some of the differences should be attributed to the discretization of the mesh, and some others to the inaccuracy of the supplied values.

CONCLUSIONS

The code adopted is able to fairly represent the actual flow, considering viscous and wall effects, though these latter have been introduced by means of empirical parameters which may somewhat affect the results. The lack of information on the geometry beyond the draft tube outlet section resulted in an inaccuracy of the schematization, as proper boundary conditions just near the recirculating flow couldn't be imposed. Despite the above described limitation, the results are realistic, and show a good capability of detecting secondary flows and overall behaviour of the investigated component.

COMPUTER RELATED QUANTITIES

computer type	ARDENT TITAN	
number of grid points	6948	
cpu time	224030	s
(cpu time)/iterations	1179	s
(cpu time)/(number of grid cells)	31.3	s
(cpu time)/(iter.* number of g.c.)	0.165	s
ratio of vectorisation	/	
parallelisation speed up	/	
convergence criterion (on u_x)	0.002	

SYMBOLOGY

ρ=density
u=velocity vector
f=body force vector
g=weight force vector
p=pressure
μ=dynamic viscosity
ν= kinematic viscosity
k=turb.kinetic energy
ϵ=viscous dissip. of k

$\hat{u}$=variation of u with respect to $\bar{u}$
$\bar{u}$=mean value of u
u_{res}=tang.vel.in viscous sublayer
u_τ=friction velocity
κ=Von Karman constant (=0.41)
E=roughness parameter (=9.0)
y=distance from the wall
Lm=mixing length

BIBLIOGRAPHY

[1] "FIDAP Theoretical Manual", F.D.I.Inc.-Evanston,Ill.,USA.

[2] SCHLICHTING, H.: "Boundary Layer Theory", McGraw Hill Book Company, New York, 1968, p. 568.

3-D-EULER FLOW ANALYSIS OF RUNNER AND DRAFT TUBE OF THE GAMM TURBINE

C. Liess
J.M.Voith GmbH
Postfach 1940, D-7920 Heidenheim, Fed.Rep.Germany

A. Ecer
Technalysis, Inc.
7120 Waldemar Drive, Indianapolis, IN 46268, USA

SUMMARY

The flow in the runner and the draft tube of the GAMM turbine was analyzed using two different types of Euler codes. The theoretical basis of the codes is outlined and the computational details are given, as well as some additional results.

INTRODUCTION

Two different Euler codes were applied to the analysis of the runner and the draft tube of the GAMM turbine:

- The runner flow was analyzed using the finite element code PASSAGE developed by Technalysis, Inc..
- The flow in the draft tube was calculated by means of a finite volume code developed by DFVLR, Göttingen, during a cooperation between Voith and DFVLR.

The two codes are described separately in the following.

ANALYSIS OF RUNNER FLOW

The runner flow solution was obtained using the three-dimensional finite element package PASSAGE. PASSAGE is capable of modeling compressible potential (irrotational, inviscid), Euler (rotational, inviscid), and Navier-Stokes (viscous) flow equations in rotating or stationary internal passages. It works well for nearly incompressible flows. In the present case, the relative inlet Mach number was 0.01.

Governing Equations

The absolute velocity vector $\underline{v}$ for rotating systems is expressed by employing a Clebsch transformation [3,4] in the form:

$$\underline{v} = \underline{\nabla}\phi + S\underline{\nabla}\eta + I\underline{\nabla}\lambda \tag{1}$$

where ϕ, S, η, I, and λ are known as the Clebsch variables [3]. Such a transformation is obtained through a variational principle which can be viewed as a generalization of the well-known Bateman's variational principle [1,8,9] used for inviscid flows. The relative velocity vector $\underline{u}$, can be obtained through the relationship:

$$\underline{u} = \underline{v} - \underline{i}_\theta \Omega r \tag{2}$$

where $\underline{i}_\theta$ is the unit vector in the tangential direction.

The reference internal energy of fluid particles, I, is a material property, and is related to the stagnation enthalpy, H_o, in the form:

$$I = H_o - \underline{v} \cdot \underline{i}_\theta \Omega r . \tag{3}$$

Euler Equations

The following five partial differential equations define the conservation property of the Clebsch variables for Euler equations:

conservation of mass: $\underline{\nabla} \cdot (\varrho \underline{u}) = 0,$ (4)

conservation of entropy: $\varrho \underline{u} \cdot \underline{\nabla} S = 0,$ (5)

conservation of η: $\varrho \underline{u} \cdot \underline{\nabla}\eta = -p/R,$ (6)

conservation of I: $\varrho \underline{u} \cdot \underline{\nabla} I = 0,$ (7)

conservation of λ: $\varrho \underline{u} \cdot \underline{\nabla}\lambda = \varrho$ (8)

where ϱ is the density, p is the static pressure, and R is the gas constant. When the velocity transformations in Equations (1) and (2) are substituted into Equation (3), the conservation of mass equation takes the form:

$$\underline{\nabla} \cdot [\varrho\underline{\nabla}\phi + \varrho S\underline{\nabla}\eta + \varrho I\underline{\nabla}\lambda] = \underline{\nabla} \cdot (\underline{i}_\theta \varrho \Omega r) . \tag{9}$$

Solution Algorithm

A block-by-block solution scheme is used for the solution of Equations (4) - (8). Rather than assembling the equations and residual vectors for the entire computational grid, the problem is described in terms of a set of blocks. The solution block structuring technique used assures exact matching of the grid points at inter-block surfaces. For obtaining a solution to the steady flow problem, the solutions between the blocks are matched in terms of all five primary variables (ϕ, S, η, I, λ) as well as the normal mass flux and the tangential velocities.

Boundary conditions

The physical boundary conditions used were the imposed flow rate, density, angular velocity, and preswirl. From this data, values of S and I at the inlet can be determined. From the inlet velocity distribution and Equation (1) the values of ϕ, η, λ are determined. At the runner outlet normal gradients of mass and entropy fluxes are set to zero.

Computational Details

The informations about non-dimensional torque, efficiency, computer, grid, and computation times are listed in the following table:

$t^* = 0.263$	
computer type	MicroVAX 3600
number of grid points	59*10*10 = 5900
CPU-time	35083 sec
CPU-time / iteration	184.6 sec
CPU-time / number of grid cells	5.95 sec
CPU-time /(iteration * no. of grid cells)	0.0313 sec

The grid structure within the runner vanes is shown in Fig. 1. The total computational domain was extended in the upstream and downstream direction.

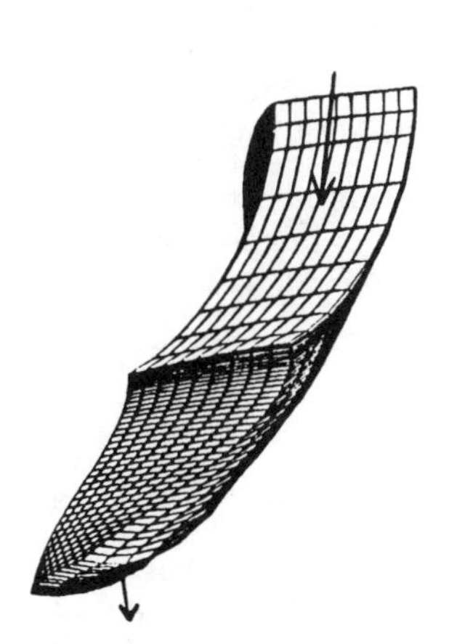

Fig. 1 Finite element grid used in the runner

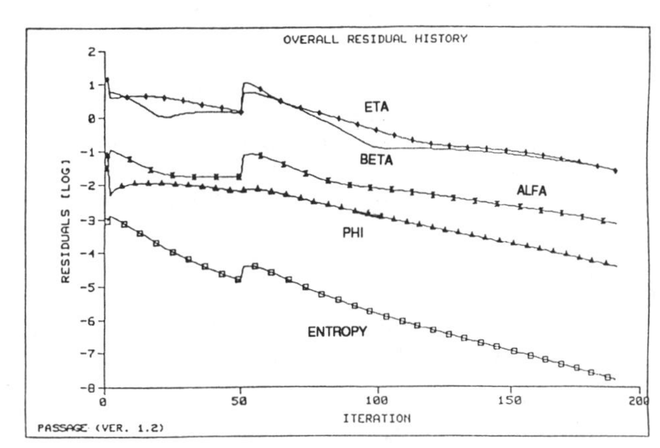

Fig. 2 Residual history

The residual history plot in Fig. 2 shows the residual history of the five equations for all iterations. The solution is considered converged when all residuals have dropped two to three orders of magnitude from their initial values.

The static pressure distributions on the runner blades in Fig. 3 show on the pressure side a relatively uniform pressure drop from the leading edge towards the trailing edge.

On the suction side a similar continuous pressure drop exists only near the crown. At the band there is a low pressure peak near the leading edge with a subsequent pressure rise towards the trailing edge.

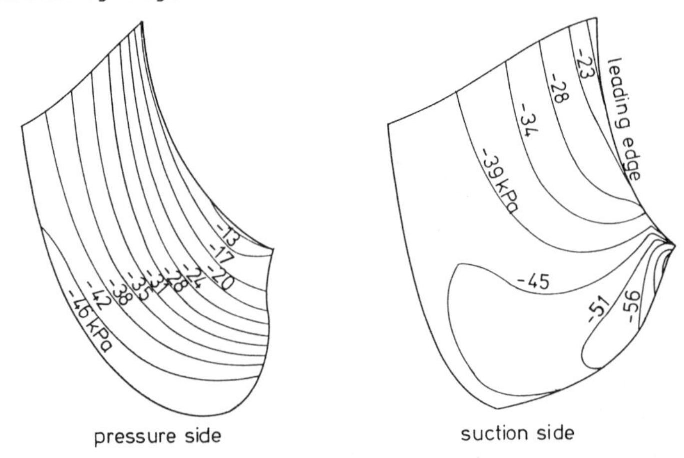

Fig. 3 Constant pressure lines on the runner blade

The computed results at the runner inlet compare well with the experiments. At the draft tube inlet there is an increasing discrepancy from the outer radius to the centre for the angle beta and the circumferential velocity component. For the other flow parameters the agreement is good, except close to the axis. These differencies can be attributed to viscous effects, which become important in the relatively large space between the runner trailing edge and the draft tube inlet.

Conclusions

The Clebsch formulation has the advantage that Potential, Euler, and Navier-Stokes flows are represented in a unified form, so that the same computer program can be used for all three types of flows. The formulation allows also the implementation of a Kutta condition for all three types of flows. The block-by-block scheme lends itself to efficient implementation on parallel computers.

ANALYSIS OF DRAFT TUBE FLOW

The real flow in the draft tube of a Francis turbine is very complicated and viscous effects are important. In order to model the flow in a reasonable way, viscous effects have to be taken into account and the Navier-Stokes equations have to be used for the analysis. The inviscid Euler code described below represents the first step in the development of a Navier-Stokes code for draft tube flows. Having in mind the limited applicability of the code, it was used anyway for the present analysis, in order to see what type of information can be deduced from such a simplified approach.

Description of the Method

The Euler equations are modified by introducing the so-called artificial compressibility [2,6]. This leads to a pseudo-instationary formulation of the continuity equation:

$$\begin{aligned}
p_t + c^2(u_x + v_y + w_z) &= 0, \\
u_t + (u^2)_x + (uv)_y + (uw)_z + p_x &= 0, \\
v_t + (uv)_x + (v^2)_y + (vw)_z + p_y &= 0, \\
w_t + (uw)_x + (vw)_y + (w^2)_z + p_z &= 0
\end{aligned} \qquad (10)$$

where p is the pressure and u,v,w are the cartesian components of the velocity vector. Equations (10) are made dimensionless by the density, a characteristic length, and a reference velocity. They can be solved by a time marching procedure, where the correct solution is obtained if a stationary state is reached. The parameter c^2 has to be chosen in suitable form, it can be either constant or variable.

Integration of the equations is done by an explicit finite volume procedure. The variables u, v, w, p are assumed to be constant in each finite volume. The system of differential equations is solved in the time direction by a four-step Runge-Kutta procedure [5].

The solution procedure is stabilized by introduction of an additional artificial viscosity. In order to improve the convergence of the method and to accelerate the solution, the maximum possible time step is chosen for each finite volume. An additional acceleration is obtained by introduction of implicit averaging of the residuals.

Boundary and Starting Conditions

In order to solve the system of ordinary differential equations resulting from the finite volume representation in the time direction, the following conditions at the boundaries as well as for starting the calculation are defined.

Along the solid walls there exists the condition of impermeability. The tangential velocity on the surfaces is determined by extrapolation from the inner flow field. The wall pressure is calculated by the momentum equation normal to the wall. At the inlet of the calculation domain the measured flow velocities are imposed and kept constant during the iterations. The pressure at the inlet is obtained by extrapolation from the inner flow field. At the outlet the components of the velocity vector are extrapolated, while a characteristic, non-reflecting boundary condition is applied for the pressure following the approach of [7] for equations (10). This condition forces the pressure at the outlet to converge to the prescribed one, once the solution itself converges. Without a given outlet pressure, however, this boundary condition has to be reduced to its characteristic part, which vanishes in case of convergence and thus can lead to convergence problems.

The calculation starts with a velocity distribution in the duct defined in such a way that the integrals of the mass flow and of the swirl are constant for all cross sections. The local pressure is determined by the Bernoulli equation, using the assumption of constant total pressure along corresponding finite volumes in the main flow direction.

Computational Details and Additional Results

The informations about the computer used for this calculation, the number of grid points, and the computing time is given in the following table:

computer type	VAX 8200
number of grid points	42*11*11 = 5082
CPU-time	45180 sec
CPU-time / iteration	45.18 sec
CPU-time / number of grid cells	11.02 sec
CPU-time /(iteration * no. of grid cells)	0.01102 sec

The grid is shown in Fig. 4. There are 11 times 11 volumes in each of the 42 cross sections. Problems of convergence of the solution, probably due to the outlet boundary condition, were encountered for higher numbers of grid cells.

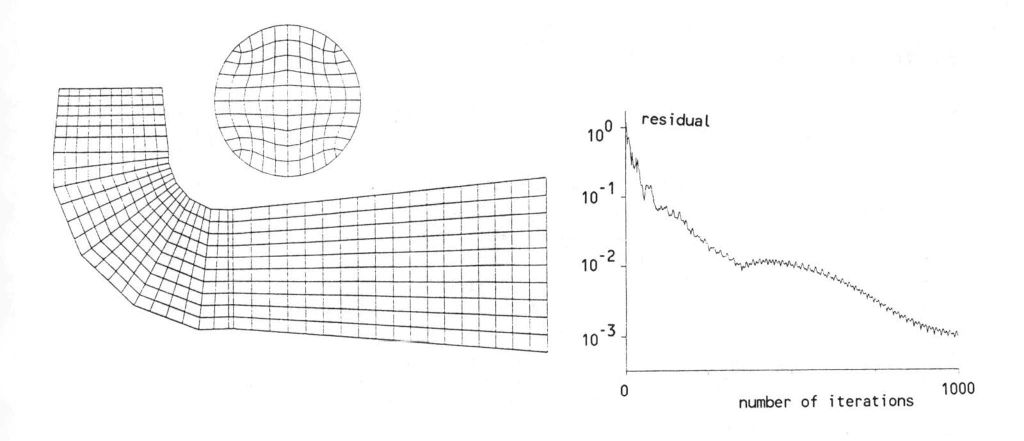

Fig. 4 Finite volume mesh used for the draft tube

Fig. 5 Residual history

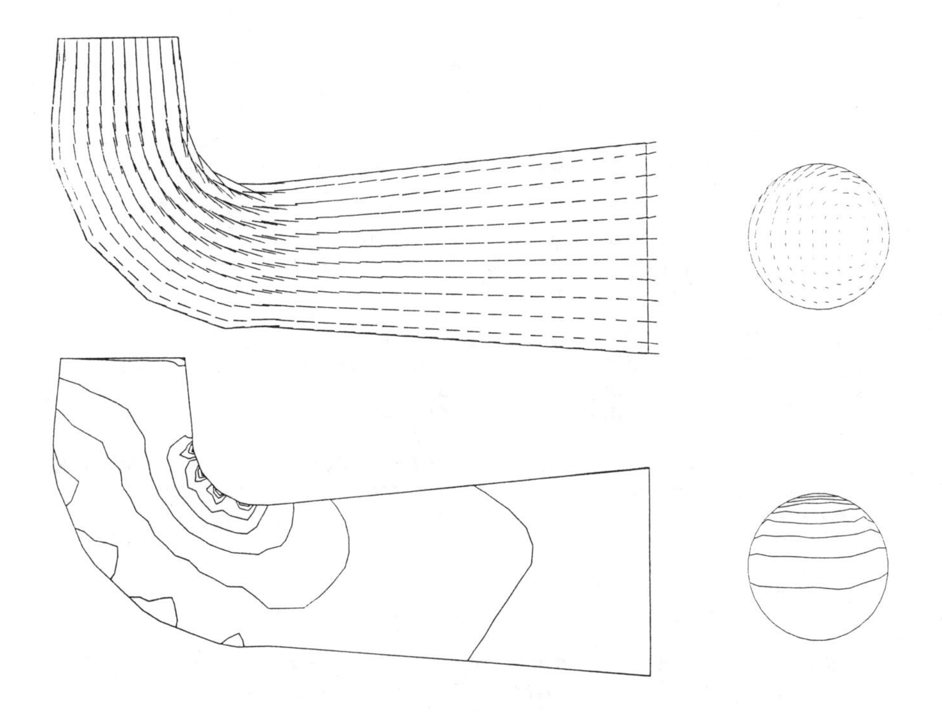

Fig. 6 Velocity vectors and constant pressure contours in the plane of symmetry and in cross section no.8, viewed in flow direction.

The development of the residual, i.e. the maximum absolute difference of velocity components in subsequent iteration steps divided by the time step, is plotted in Fig. 5 against the number of iterations. The calculation was stopped when the residual dropped below a value of 10^{-3}.

Fig. 6 shows velocity vectors and constant pressure lines in the plane of symmetry and in cross section no.8 of the draft tube. The present inviscid analysis shows no separation downstream of the elbow, in contrast to most of the viscous calculations. However, the pressure development along the draft tube looks reasonable and similar to some of the viscous results. It would be very interesting to compare it to measurements, in order to assess the capabilities of an inviscid computation for such flow conditions.

REFERENCES

[1] BATEMAN, H.: "Notes on a differential equation which occurs in the two-dimensional motion of a compressible fluid and the associated variational problems", Proc. Royal Society, Sec. A, London, 125 (1929), pp. 598-618.

[2] CHORIN, A. J.: "A numerical method for solving incompressible viscous flow problems", J. Comp. Physics 2 (1967), pp. 12-26.

[3] CLEBSCH, A.: "Über eine allgemeine Transformation der hydrodynamischen Gleichungen", Crelles J. Reine Angew. Math., 56 (1859), pp. 1-6.

[4] ECER, A., AKAY, H. U.,SHEU, W. H.: "Variational finite element formulation for viscous compressible flows", Symp.Num.Meth.Compressible Flows, ASME Winter Meeting, Anaheim, Cal., Dec. 7-12,1986, AMD-Vol. 76, pp. 5-17.

[5] JAMESON, A., SCHMIDT, W.: "Recent development in finite-volume time-dependent techniques for two and three-dimensional transonic flows", VKI Lecture Series 1982-04.

[6] PEYRET,R., TAYLOR,T.D.: "Computational methods for fluid flow", Springer Verlag, New York, 1983.

[7] RUDY, D. M.,STRIKWERDA, J. C.: "A non-reflecting outflow boundary condition for subsonic Navier Stokes calculations", J. Comp. Physics 36 (1980), pp. 55-70.

[8] SELIGER, R.L., WHITMAN, G.B.: "Variational priciples in continuum mechanics", Proc. Royal Society, Sec. A, London, 305 (1968), pp. 1-25.

[9] SERRIN, J.: "Mathematical principles of classical mechanics", Handbuch der Physik, Vol.VIII/I (1959), Springer Verlag.

A POTENTIAL PREDICTION OF THREE-DIMENSIONAL INCOMPRESSIBLE FLOWS THROUGH TURBOMACHINERY BLADE ROWS

N. Lymberopoulos, K. Giannakoglou, P. Chaviaropoulos,
K.D. Papailiou
Lab of Thermal Turbomachines
National Technical University of Athens
P.O. Box 64069, 157 10 Athens, Greece.

SUMMARY

A method is presented for the numerical solution of incompressible three-dimensional flows in turbomachinery blade rows. The flow is assumed to be irrotational in the absolute frame of reference.

The governing equations and their boundary conditions are presented and discussed. Centered finite-difference/ finite volume schemes are used for the discretization of the governing equations, after a body-fitted coordinate transformation. The resulting linear algebraic system of equations is inverted by a fast iterative procedure. Numerical results are presented for the 1989 Gamm Workshop Francis Runner geometry.

INTRODUCTION

Considering the flow to be irrotational in the absolute frame of reference, this method solves an elliptic type equation for the velocity potential Φ on a computational domain, which is provided after a body-fitted coordinate transformation. Special care has been taken for the implementation of the periodicity and the non-linear Kutta condition.

A preconditioned minimization scheme is adopted for the inversion of the characteristic matrix, based on the combination of the MSIP [5] approximate factorization technique with the linear GMRES [4] algorithm. This method was proved to be fast and accurate in several applications [1], [2].

GOVERNING EQUATIONS

Assuming the flow to be irrotational in the absolute frame, the absolute velocity ($\vec{V}$) is defined from the velocity potential (Φ), $\vec{V}=\nabla\Phi$. The relative velocity ($\vec{W}$) is expressed as

$$\vec{W} = \vec{V}-\vec{U} = \vec{V}-\vec{\omega}x\vec{r}. \qquad (1)$$

The continuity equation states that

$$\nabla\cdot\vec{W}=\nabla\cdot(\nabla\Phi-\vec{U})=0. \qquad (2)$$

The relative stagnation pressure p_{tR} is defined as

$$p_{tR} = p + \frac{1}{2}\rho W^2 - \frac{1}{2}\rho U^2 + \rho g z \qquad (3)$$

where ρ is the density and p is the pressure. Imposing a vorticity free velocity distribution at the inlet, P_{tR} is taken to be constant in the whole field. Once the velocity field is known, eq.3 may be used for the direct computation of the pressure field. On the solid boundaries $W_n=\vec{W}\cdot\vec{n}=0$, while at the inlet and exit planes W_n is specified so as to respect the integral continuity equation. The circulation (Γ) around the blade is specified from the Kutta condition ($p^+=p^-$) and provides the value for the jump in potential $\Phi^+-\Phi^-$ at all points along the trailing edge.

COORDINATE TRANSFORMATION AND BOUNDARY CONDITIONS

A transformation has been used to pass from an arbitrary flow region in the physical space (x_i ; i=1,2,3) into an orthogonal parallelepiped with equidistant nodes placed in the computational space (u_j ; j=1,2,3). The topological aspects of the coordinate transformation are presented in Figure 1 for a simplified machine configuration.

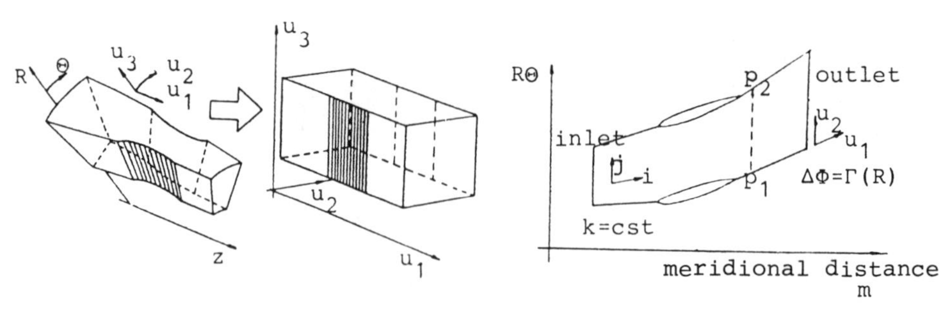

Fig. 1 Coordinate transformation

Fig. 2 Periodicity

The governing equations, after the coordinate transformation, are written in the following form

$$\partial(JW^i)/\partial u_i = 0, \tag{4}$$

$$W^i = g^{ij}W_j = g^{ij}(\partial\Phi/\partial u_j - U_j) \tag{5}$$

where W^i, W_j ; $i,j=1,2,3$ are, respectively, the contravariant and covariant relative velocity components and g^{ij} stands for the contravariant metrics tensor.

For the configuration given in Fig.1 the following boundary conditions are imposed on the scalar potential Φ

- on the blade surface $W^2=0$
- on the inlet and exit planes W^1=given
- on the crown and band planes $W^3=0$.

The periodicity conditions take the form (see Fig.2)

$$\Phi_{p2} = \Phi_{p1} + \Delta\Phi(R),$$

$$\left.\frac{\partial\Phi}{\partial u_2}\right|_{p2} = \left.\frac{\partial\Phi}{\partial u_1}\right|_{p1}, \quad \left.\frac{\partial(\Delta\Phi)}{\partial u_1}\right|_{u_2=u_{2max}} = 0. \tag{6}$$

On the upstream periodic boundary the $\Delta\Phi$ value is a constant which is computed on the inlet plane through mean peripheral integration of the tangential $\vec{V}$ component. On the downstream periodic boundary the $\Delta\Phi$ values, representing the local circulation, vary with R. The local value of $\Delta\Phi(R)=\Gamma(R)$ is provided by the Kutta condition

$$|W|^{+}_{TE} = |W|^{-}_{TE}, \tag{7}$$

which introduces a non-linearity in the problem. Since the value of $(\Delta\Phi)$ varies in general along the trailing edge line, there is a jump of the contravariant component W^3 along the vortex sheet $(W^3|_{p2} \neq W^3|_{p1})$. The problem was solved in such a way that the periodicity conditions were not assured for the W^3-velocity component.

DISCRETIZATION AND SOLUTION PROCEDURE

Equation (6) is discretized by second order accurate finite-difference/finite volume schemes, which provide the flux-balance in the transformed grid cells [3]. All Neumann-type boundary conditions on (Φ) are treated in an implicit way through the flux-balance procedure. The resulting system of algebraic equations takes the form

$$[A]\{\Delta\Phi\} = \{q\},$$
$$\{\Phi\}^{n+1} = \{\Phi\}^{n} + \{\Delta\Phi\} \,, \tag{8}$$

which is solved iteratively for the pseudo-time-step (n). [A] is a 19-diagonal non-symmetric matrix, depending only on the coordinate transformation metrics, and {q} is the vector containing the fluxes.

In order to invert matrix [A], an approximate L-U factorization scheme of the 19-diagonal MSIP version [5] is performed first, providing the lower and upper triangular diagonal matrices [L] and [U] of the same structure as [A]. Then, the first part of eq.(8) is solved for $\{\Delta\Phi\}$ in the following preconditioned form:

$$[U^{-1}][L^{-1}](\{q\}-[A]\{\Delta\Phi\}) = 0. \tag{9}$$

The above system is solved by the linear GMRES (m) technique [4]. At each pseudo-time step of the iterative procedure, the Kutta condition (equation (7)) has to be satisfied. The Kutta condition forms a non-linear system of equations which is solved for $\Gamma(R)$ along the trailing edge by a non-linear GMRES algorithm. The initial estimation of $\Gamma^*(R)$ is provided at each time step by a linear approximation of the Kutta condition

$$W_{TE}^{1+} = W_{TE}^{1-} \tag{10}$$

which leads to the solution of a tridiagonal matrix for $\Gamma^*(R)$.

GRID GENERATION

The above described method was applied to a Francis Runner presented in the 1989 Gamm Workshop. A two-dimensional (70x17) H-type grid was generated on the meridional plane (z,R) (see Figure 3) using a tranfinite interpolation procedure. This meridional grid respects the grid points originally given which correspond to the blade section. However, in order to avoid the singular line which occurs (due to the coordinate transformation) after the nacelle of the machine, the crown geometry was modified as shown in Figure 3. Because of the potential formulation of the problem the authors believe that this modification does not effect the flow in the near-blade region. Once the meridional grid was formed, a three-dimensional (70x17x10) grid was obtained through peripheral stacking with constant peripheral angle (theta) increments. The theta angle distribution on the periodic parts of the meridional plane was imposed in a way that respects its slope in the leading and trailing edge regions. A perspective view of the three-dimensional calculation domain is shown in Figure 4.

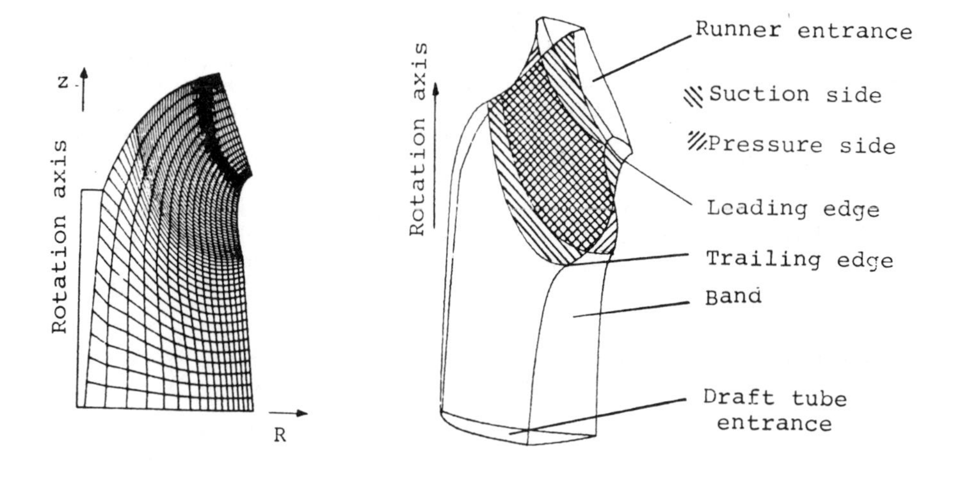

Fig. 3 Modified meridional grid

Fig. 4 Contour of calculation domain

RESULTS AND DISCUSSION

The method presented above was applied on the (70x17x10) computational grid described in the previous paragraph. In order to take into account the given experimental data, the boundary conditions were specified in the following way:
(a) the $\vec{W}\vec{n}=W_n$ distribution on the inlet and exit planes was specified from the experimental measurements. The flow rate that was computed on the inlet plane was $V=.346\ m^3/s$, while the nominal flow rate was given to be $V^*=0.372\ m^3/s$, (b) the $\Delta\Phi$ value which is imposed on the upstream periodic boundary was computed after a mean integration on the inlet-plane of the experimental absolute velocity in the peripheral direction, (c) the normal velocity distribution at the exit plane was modified in order to respect the inlet flow rate, (d) the imposed angular velocity of the rotor was the given $\omega=52.36$ rad/s.

It should be noted at this point that due to the assumption of potential flow, the imposed inlet boundary conditions were not in exact agreement with the given experimental data. Specifically, the vorticity-free assumption calls for a uniform pre-swirl (RV_u) distribution and a uniform relative stagnation pressure distribution on the flow inlet plane. Consequently, the experimental data were described as best as possible. The inability of simulating the incoming vorticity effects is the major limitation of the potential flow solvers compared with the full Euler ones.

Figure 5 presents the convergence history of the method as a plot of the maximum residual of equation (9) in respect to the GMRES(m=4) iteration. The computational cost for a fast PC-386 machine and in a single precision environment was 45 CPU minutes for 30 GMRES(4) iterations.

Once the velocity field was computed, the pressure field was obtained from equation (3). The non-dimensional pressure coefficient c_p^* was, then, computed from the relation

$$c_p^* = (p-p_{ref})/\rho E \tag{11}$$

where p_{ref}=94.300 N/m² is a given reference static pressure, ρ=1000 kg/m³ is the water density and E=58.42 J/kg is the specific hydraulic energy. The computed c_p^* contours on the blade suction and pressure sides are shown in Figure 6. The exit flow angle distribution was not accurately predicted since neither the incoming vorticity nor the flow viscosity effects (flow losses) were taken into account.

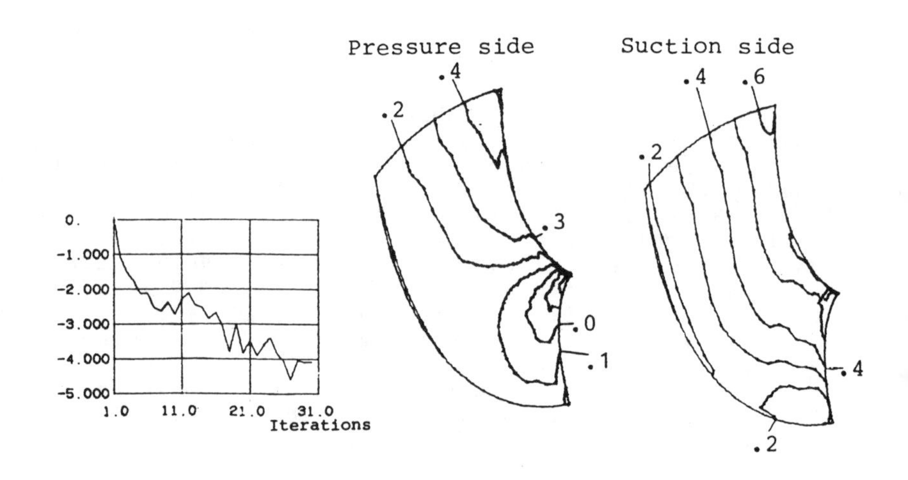

Fig. 5 Convergence of GMRES

Fig. 6 Lines of equal c_p

It is expected that, when the theoretical results will be compared to experiment, differences will be present, as this is the case with the exit flow angle. However, since the incoming vorticity is not particularly important, it is expected, as well, that the computed results will not deviate excessively from the measured ones.

CONCLUSIONS

A method for the numerical simulation of incompressible three-dimensional flows in turbomachinery blade rows was presented. The method is a finite-difference/finite-volume potential solver and makes use of an iterative scheme based on the MSIP preconditioned GMRES algorithm. The technique has been applied to the Francis Runner geometry presented in the 1989 Gamm Workshop. The inlet and exit flow boundary conditions needed for the present computation were obtained from existing experimental data, after suitable modifications due to the limitations of the potential flow hypothesis. Although the inlet flow vorticity effects were not taken into account in this formulation, the predicted blade loading is in reasonable agreement with the experimental results.

REFERENCES

[1] CHAVIAROPOULOS, P., PAPAILIOU, K.D.: "A Full potential prediction of a HAWT rotor performance", Proc. of 1988 European Community Wind Energy Conference, Herning, Denmark, 1988.

[2] GIANNAKOGLOU, K., CHAVIAROPOULOS, P., PAPAILIOU, K.D.: "Acceleration of standard full-potential and elliptic Euler solvers, using preconditioned generalized minimal residual techniques", Proc. 1988 ASME Conf., Chicago 1988.

[3] HOLST, T.L.: "Numerical computation of transonic flow governed by the full potential equation", VKI LS 1983-04, Computational Fluid Dynamics, March 1983.

[4] SAAD, Y., SCHULTZ, M.M.: "GMRES: A Generalized minimal residual algorithm for solving non-symmetric linear systems", Res. Rep. YALEU/DCS/RR-254, Aug. 1983.

[5] ZEDAN, M., SCHNEIDER, G.E.: "A Three-dimensional modified strongly implicit procedure for heat conduction", AIAA Journal, 21 (1983), No 2, pp. 295-303.

3-D FLOW ANALYSIS IN THE RUNNER AND THE DISTRIBUTOR OF FRANCIS TURBINES

T. Nagafuji and T. Suzuki
Toshiba Corporation
20-1 Kansei-cho, Tsurumi-ku, Yokohama, Japan

T. Kobayashi and N. Taniguchi
Institute of Industrial Science, The University of Tokyo
7-22-1 Roppongi, Minato-ku, Tokyo, Japan

SUMMARY

The three-dimensional computation of incompressible internal flows was carried out for the runner and the distributor of the GAMM Francis turbine. The runner is calculated as potential flow using the finite element method and the distributor as viscous flow using the finite volume method.

INTRODUCTION

The numerical simulation in steady and incompressible flow fields is in progress from inviscid to viscous flows.

In case of inviscid flow, the potential equation has widely been used because it has the advantages of easy handling and low computational cost, which make easier to use it as a design tool, while the Euler equations, which allow for the transport of vorticity in the calculation, have been studied as a recent technique.

On the other hand, the Navier-Stokes equations with turbulence modeling have been studied to solve viscous flow in turbomachinery. However, it is limited to low Reynolds number within laminar flow region because the present technique includes some problems to be solved for the turbulent flow at higher Reynolds numbers.

This paper shows the numerical results of two examples, one of which is the potential flow analysis in the runner, while the other is the viscous flow analysis in the distributor under laminar flow condition.

FLOW ANALYSIS IN THE RUNNER

Finite element method

The finite element method [1] is used to discretize the

governing equations. The basic equations of potential flow in a rotating runner are as follows:

$$\frac{1}{2}(\nabla\phi)^2 + (\boldsymbol{\omega}\times\mathbf{r})\cdot\nabla\phi + U + \frac{p}{\rho} = \text{const.}, \tag{1}$$

$$\nabla^2\phi = 0, \tag{2}$$

where ϕ is the velocity potential function, U is the potential of an external force, ρ is the constant density of water, p is the static pressure, $\boldsymbol{\omega}$ is the angular velocity vector, $\mathbf{r}$ is the distance vector from the axis of rotation.

The procedure of discretizing the basic equations directly into finite element equations has been reported in reference [2]. In this study, the Cartesian coordinates and hexahedral elements are used. The solution of the potential flow can be easily obtained by the process of direct matrix inversion.

Numerical modeling and boundary conditions

The global domain of the runner for the calculations is composed of the geometry between the measuring sections at inlet and outlet which were given by the organizers. Fig. 1 shows the result of mesh discretization in which the number of grid points is 4284 (36 in the meridional direction, 17 along the height and 7 in the circumferential direction).

The domain in the circumferential direction is limited to a blade-to-blade region. Therefore, the periodic condition is introduced in the vaneless spaces of inlet and outlet passages. The circulation has a constant value in each space as free vortex flow is assumed at the inlet and the outlet. The experimental data at the outlet shows nearly zero circulation from center to crown side. Therefore, it is assumed that the circulation is zero at the outlet. On the other hand, the average value of experimental data is used as the circulation at the inlet.

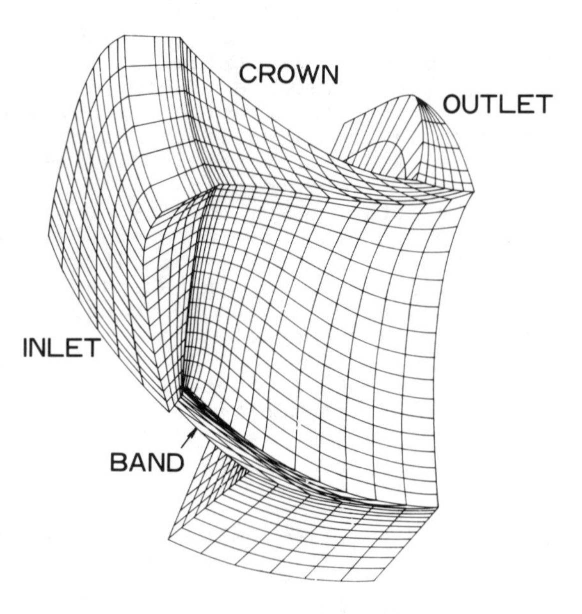

Fig. 1 Finite element model of runner flow domain

Another boundary condition is the normal velocities at the inlet and outlet sections. These values are determined by using the experimental data for each element in the height at the inlet and in the radial direction at the outlet, and are slightly corrected so that the mass flow across them corresponds to the given value.

The relative entire flow in the rotating runner is finally solved by using the given rotational speed.

Numerical results

The computer related quantities which were used on the calculation are summarized on Table 1.

Table 1 Computer related quantities

Computer type	---	ACOS 1000 (7MFLOPS)
CPU time	---	290 sec (4min 40sec)
(CPU time)/(No. of grid points)	---	0.0677 sec

Note; No iteration, vectorization and parallelisation

As described previously, the numerical simulation is carried out by using the averaged circulations which are derived from the experimental data at the inlet and the outlet. Therefore, the inlet and outlet boundary conditions do not coincide exactly with the experimental ones. These differencies give a certain error for the numerical result regardless of viscous effects.

The relative velocity fields on the blade surfaces and on the blade-to-blade surfaces at the crown, center and band sections are drawn on Figs. 2 and 3.

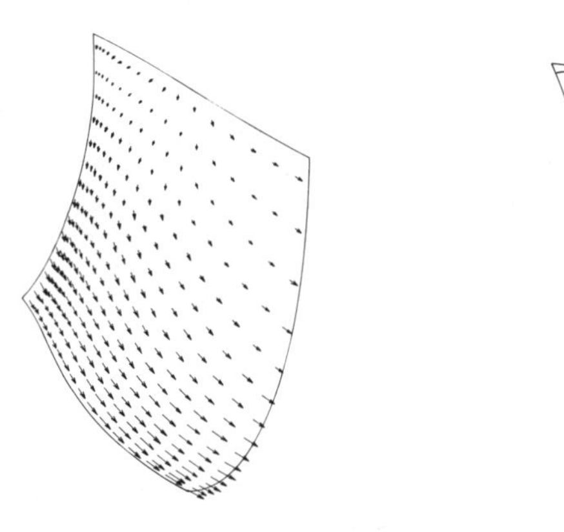

(a) Pressure side (b) Suction side

Fig. 2 Flow patterns on blade surfaces

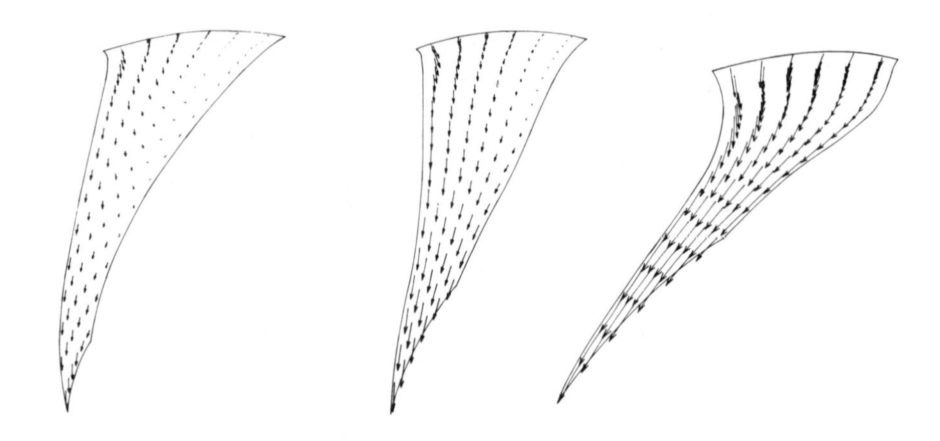

(a) Crown section (b) Center section (c) Band section

Fig. 3 Flow patterns on blade-to-blade surfaces

The global physical quantities which are obtained by the calculation are as follows:

Nondimensional torque, $t^* = 0.2937$,

Efficiency, $\eta = 1.039$.

FLOW ANALYSIS IN THE DISTRIBUTOR

Finite volume method in the body-fitted grid

The finite volume method [4] is used to discretize the governing equations in the body-fitted grid. The equations are integrated in the generic control cell shown in Fig. 4 which appears along the nonorthogonal coordinate system. The staggered technique is adopted for the velocity components, which are defined as the projections on the coordinates. The generalized form which is derived from the conservation equations is,

$$\iiint_{Vol} \frac{\partial \phi}{\partial t}\, dV + \iint_{Anb} (\mathbf{J}_{nb} \cdot d\mathbf{n}_{nb}) = \iiint_{Vol} S\, dV \qquad (3)$$

where $\mathbf{J}(=\mathbf{v}\phi - \Gamma\nabla\phi)$ is the total flux on the cell surface, $\mathbf{n}$ is the unit face vector, ϕ is a general variable(a scalar or a component of vector), $\mathbf{v}$ is the velocity vector, Γ is the diffusive coefficient, S is a source of ϕ and the suffix nb means e, w, n, s, t and b as shown on Fig. 4.

The surface integration is estimated in the following way. The example on the surface "e" is,

$$\iint_{Ae}(\mathbf{J}_e \cdot d\mathbf{n}_e) = JD_e F(|Pe_e|)(\phi_e - \phi_p) + JM_e \phi_{UP} \quad \ldots\ldots(4)$$

with

$$JM = \mathbf{v} \cdot \mathbf{A}, \quad JD = (\Gamma/PE)(\mathbf{e} \cdot \mathbf{A})$$

$$Pe = JM/JD \quad \ldots\ldots(5)$$

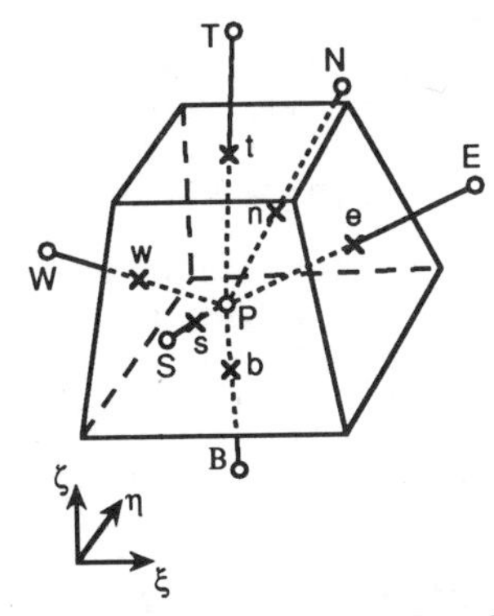

Fig. 4 Control cell of a point 'P' and its neighboring points

where Pe is the Peclet number, **e** is the unit vector along the ξ coordinate axis, **A** is the face vector, PE is the segment length from P to E and ϕ_{UP} means upwind value. It is noticed that both the velocity and the unit vectors are defined in the direction of the segment P to E, but are not generally normal to the face vector.

The hybrid scheme is adopted in this study, where the function F is switched at Pe=2 from a central differencing formulation to an upwind scheme to reduce dispersion errors.

This method has the advantage that the above formulation can be derived directly from the equation in the physical domain, i.e. it is not transformed into any other coordinate. The staggered technique effectively suppresses the numerical oscillation though the formulation becomes complicate. For example, the velocity on the cell face is expressed with the one component value which is defined there and the interpolation values of the other two components in the neighboring positions in the interpolations of flux.

Concerning the continuity equation, a SIMPLE algorithm [3] can be as valid in the body-fitted grid as in the squared grid along the Cartesian coordinates in case that the components or velocity are nearly normal to the cell faces, where angles larger than about 60 degrees are expected.

At first, the discretized form of the continuity equation is shown as follows:

$$\Sigma JM_{NB} = 0 \qquad (NB: E, W, N, S, T \text{ and } B) \qquad (6)$$

where JM is mass flux on the cell face defined in eq.(5). According to the conception of S.V.Patankar [3], the residual pressure and velocity (expressed by p' or u') are splitted from the assumptions of p or u. After the mass flux(JM) is also divided into the assumptive and the residual values, eq.(6) is modified with the relations of the residual pressure and velocity derived from the momentum equation as follows:

$$\Sigma JM'(p') = -\Sigma JM(u,v,w) \qquad (6')$$

where JM' is the residual flux which is a linear function of p'. The assumed values are corrected by the following equations with a relaxation factor (α):

$$u = u + f(p'), \quad p = p + \alpha p'. \tag{7}$$

The present method has the same procedure as the original one in the Cartesian coordinates though it reduces the relaxation factors of the pressure correction and the momentum equations. The convergence of result can be estimated by the residual pressure(p') which is usually less than 0.001 in the ratio of the p' to the range of pressure.

Numerical modeling and boundary conditions

The steady viscous flow is simulated by the above numerical method. Instead of turbulence models, the viscosity is defined as 100 times the physical value in the whole region in order to consider the eddy (turbulent) viscous effect. The Reynolds number which is defined by the average inlet velocity and the height between the upper and lower rings is 1000. The time derivative term is neglected under the assumption of steady flow.

The three-dimensional body-fitted grid which was generated by an algebraic method is shown on Fig. 5. The number of grid points is 67,500(75 in the radial direction, 30

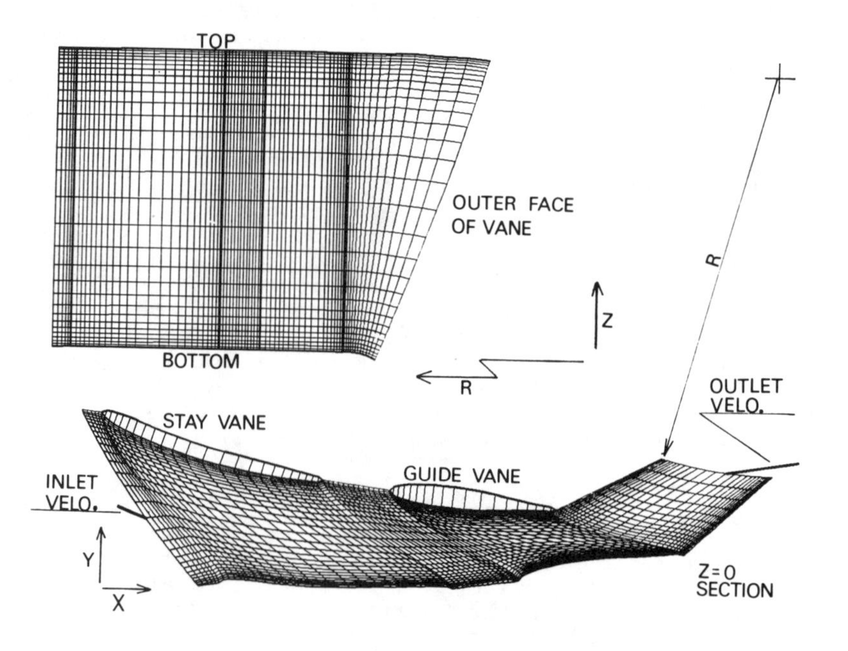

Fig. 5 Grid configuration of distributor flow domain

along the height and 30 in the circumferential direction). No extra technique is used for the orthogonality grid.

The calculating domain is composed of the one sector with the angle of 15 degrees just between the vanes. Therefore, the periodic condition is imposed in the circumferential vaneless spaces.

It is necessary to impose either the residual pressure (p') in the boundary control volume or the normal velocity at the boundary face for the p' of eq.(6'). In order to equal to zero p', the experimental data given by the organizers are used for the velocity at the inlet and the pressure at the outlet as the boundary conditions on the calculation. The streamwise derivative is set to zero at the outlet. On the upper and lower rings and on the surfaces of stay and guide vanes, the slip condition is imposed in order to simulate the flow in higher Reynolds number.

Numerical results

The computer related quantities which were used for the calculation are summarized on Table 2.

Table 2 Computer related quantities

Computer type	- HITAC S820/80 (2GFLOPS)
CPU time	- 505 sec (8min 25sec)
(CPU time)/Iteration	- 0.84 sec
(CPU time)/(No. of grid points)	- 0.0075 sec
(CPU time)/(Iteration x No. of grid points)	- 1.25×10^{-5} sec
Ratio of vectorization	- over 20 times
(Vectorized CPU time)/(Total CPU time)	= 85 %
Parallelisation	- no applied

The calculation of 600 steps is performed in the SIMPLE algorithm. The convergence condition can almost be satisfied as shown on the diagram of Fig. 6, where P' and (Pmax-Pmin) mean the pressure correction per step and the maximum difference of pressure distribution in the domain, respectively. The array super-computer attains the high acceleration ratio by vectorization which is estimated over 20 times as the scalar computation.

The velocity fields are drawn on Fig. 7 on the constant height sections Z=-50 mm, Z=0 mm and Z=50 mm and on the

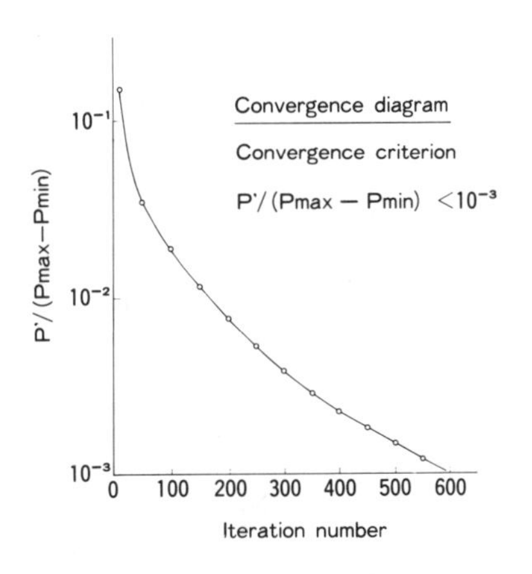

Fig. 6 Convergence diagram

constant radial sections of nearly the trailing edges of vanes. The differences among the height sections can be estimated to be small, so that the case could conceivably be simulated with the two dimensional assumptions, though some secondary flows are recognized on the radial sections.

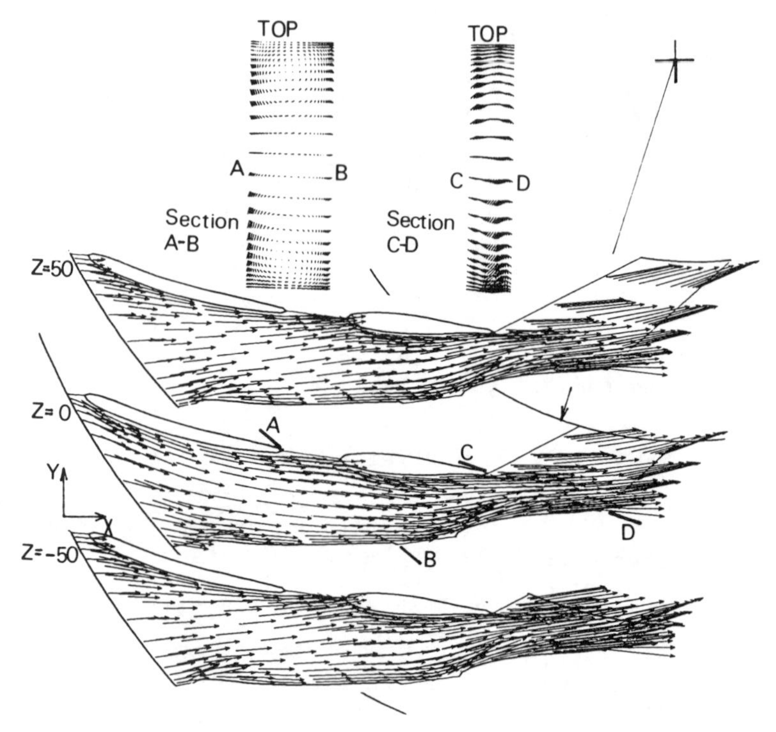

Fig. 7 Flow patterns on horizontal and vertical sections

REFERENCES

[1] CHUNG, T. J.:"Finite element analysis in fluid dynamics", McGraw-Hill, New York 1978.

[2] NAGAFUJI, T., MORII, H.:"A flow study in Francis turbine runner", Proc. of 10th IAHR symposium, Tokyo, 1980, pp. 583-594.

[3] PATANKAR, S. V.:"Numerical heat transfer and fluid flow", Hemisphere Pub. Co., New York 1980, pp. 113-135.

[4] TANIGUCHI, N., ARAKAWA, C., KOBAYASHI, T. and TAGORI, T.: "Numerical simulation of the flow around a car with finite volume method on general co-ordinate system", Trans. JSME (B), 55 (1989), No.518, pp. 3026-3032 (in Japanese).

CALCULATION OF THE DRAFT TUBE GEOMETRY USING THE FINITE-ELEMENT-CODE FENFLOSS

A. Ruprecht
Universität Stuttgart, Institut für Hydraulische Strömungsmaschinen
Pfaffenwaldring 10, D-7000 Stuttgart 80, Germany

SUMMARY

The calculation of the draft tube geometry has been carried out, using the finite element program FENFLOSS. This program solves the averaged Navier-Stokes equations. To describe the turbulence behaviour the k-ε model of turbulence has been used. The calculated velocity field in the draft tube shows a large region of recirculation.

INTRODUCTION

The draft tube part of the workshop specification is investigated. For the calculation a turbulent viscous flow is taken into account. Therefore the averaged Navier-Stokes equations are considered. For evaluating the state of turbulence the eddy viscosity hypothesis of Boussinesq is assumed. The eddy viscosity ν_t is calculated from the k-ε model of turbulence.

The calculation is carried out using the finite element program FENFLOSS [6], which is designed for incompressible flows in two- and three-dimensional geometries. In order to reduce the required memory, which otherwise is too large for 3D problems, FENFLOSS uses a sequential algorithm for the solution of the necessary equations.

BASIC EQUATIONS

The steady state flow of an incompressible fluid is considered. Therefore the averaged Navier-Stokes equations combined with the eddy viscosity concept can be expressed as

$$\rho U_j \frac{\partial U_i}{\partial x_j} = - \frac{\partial P}{\partial x_i} + \frac{\partial}{\partial x_j} \left[\rho (\nu + \nu_t) \left(\frac{\partial U_i}{\partial x_j} + \frac{\partial U_j}{\partial x_i} \right) \right] + \rho\, g_i . \tag{1}$$

Here and in the following the tensor notation is used. For the continuity equation one obtains

$$\frac{\partial U_i}{\partial x_i} = 0 . \tag{2}$$

The eddy viscosity ν_t is calculated from

$$\nu_t = c_\mu \frac{k^2}{\varepsilon} \quad . \tag{3}$$

The turbulent kinetic energy k is obtained from the transport equation

$$U_i \frac{\partial k}{\partial x_i} - \frac{\partial}{\partial x_i} \left[\left(\nu + \frac{\nu_t}{\sigma_k} \right) \frac{\partial k}{\partial x_i} \right] = G - \varepsilon \tag{4}$$

with the production term G

$$G = \nu_t \left(\frac{\partial U_i}{\partial x_j} + \frac{\partial U_j}{\partial x_i} \right) \frac{\partial U_i}{\partial x_j} \tag{5}$$

and the dissipation rate ε from

$$U_i \frac{\partial \varepsilon}{\partial x_i} - \frac{\partial}{\partial x_i} \left[\left(\nu + \frac{\nu_t}{\sigma_\varepsilon} \right) \frac{\partial \varepsilon}{\partial x_i} \right] = c_{1\varepsilon} \frac{\varepsilon}{k} G - c_{2\varepsilon} \frac{\varepsilon^2}{k} \quad . \tag{6}$$

For the model constants the standard values suggested, e. g. in [4], have been used. These values are summarized in tab. 1.

Tab. 1: Constants of the k-ε model

c_μ	σ_k	σ_ε	$c_{1\varepsilon}$	$c_{2\varepsilon}$
0.09	1.0	1.3	1.44	1.92

The high-Reynolds-number form of the k-ε model (eqs. (3-6)) is known to produce quite poor results in the presence of swirl or strong streamline curvature. Accordingly, there are different modifications in the literature (e.g. [1,3]) to improve the performance in these kind of flows. However, the standard model has been chosen for the calculation in this paper, since the modifications are not well tested for 3D separated flows.

PROGRAM DESCRIPTION

FENFLOSS is a finite-element code for the calculation of the turbulent flow of an incompressible fluid. It can handle two- and three-dimensional problems. In its present version it is restricted to steady state flows. It uses either the Standard-Galerkin formulation, which leads to a symmetrical discretisation, or a Petrov-Galerkin procedure, leading to a kind of skewed upwinding due to an unsymmetric weighting function. Different two-equation models of turbulence are embodied, among them the k-ε model.

FENFLOSS uses a sequential algorithm for solving the different equations in order to reduce the memory requirement. This algorithm is shortly explained in the following. For further details the reader is referred to [6].

The iteration algorithm, schematically shown in fig. 1, is divided into two parts. In the first part the momentum and continuity equations are solved. This is done in an iterative manner, processing sequentially one momentum equation after the other. The pressure is calculated from the continuity equation by the Uzawa algorithm described in [7]. In order to ensure continuity a penalty-like term, multiple of the continuity equation, is added to the momentum equations. This procedure is executed in a local iteration loop without reassembling the system matrices, until the continuity equation is satisfied up to required accuracy.

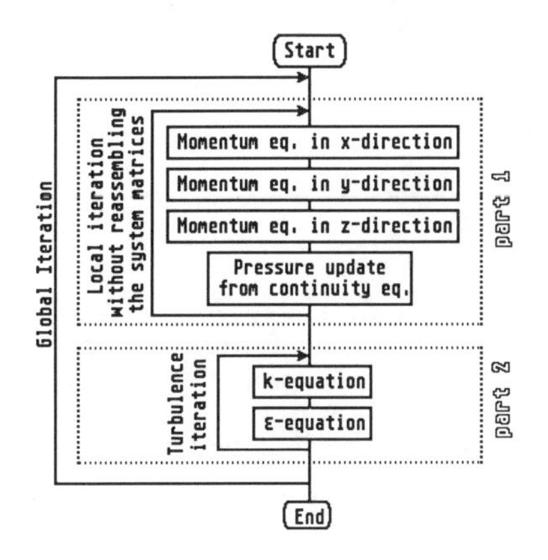

Fig. 1: Iteration scheme

In the second part, the turbulence equations (k and ε) are solved. This is also done sequentially and a few, usually two or three, iteration steps are performed.

These two parts are carried out in a global iterative loop until convergence is obtained.

All convergence information and computer time requirements shown later are related to this global iteration.

The advantage of this sequential procedure is that it requires less memory than the simultaneous handling of all the momentum and continuity equations and it can easily be processed in parallel also on local memory machines.

GRID

The flow domain is divided into isoparametric eight-node brick elements. On these elements a linear approximation for the velocity components and the turbulence quantities (k and ε) is enforced, while the pressure is kept constant on the element.

The finite-element mesh has been created by an automatic generator. For that purpose the draft tube has been cut into 43 disks of circular cross-section and each disk has been divided into 288 elements. The mesh used is shown in fig. 2. It consists of 13772 nodes and 12384 elements.

BOUNDARY CONDITIONS

The boundary conditions used are shown in fig. 3. At the inlet the measured velocity components are specified. Since no information is available for the turbulence quantities the turbulent kinetic energy k is set to a constant value, which approximately corresponds to a fully developed pipe flow. The dissipation rate ε is calculated using the mixing length formula of Nikuradse [2]. These settings of the turbulence quantities are quite arbitrary.

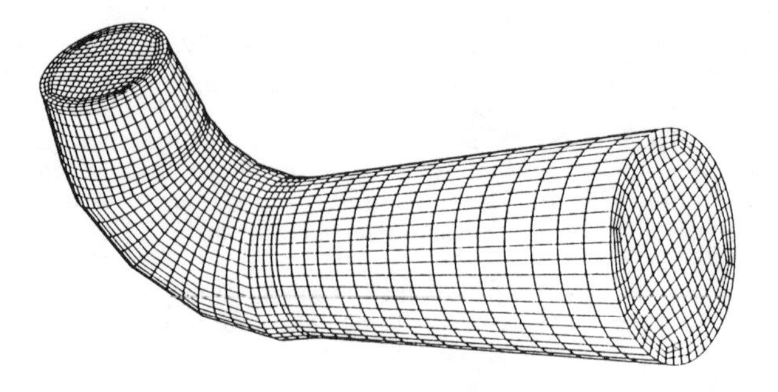

Fig. 2: Finite-element mesh

A wall function approach is employed to treat the wall boundary conditions since the k-ε model applied is valid only in the fully turbulent region and, therefore, can not describe the behaviour near the wall. These wall functions are not very suitable for separated flows. However, the use of a low-Reynolds-number k-ε model is very expensive since it requires a very fine grid in the near wall region. Even the use of a one-equation model for the near wall region, as proposed by Rodi [5], seems to be very costly for 3D flows. Therefore the wall function approach has been chosen. For implementation details the reader is referred to [6].

Since the boundary conditions at the outlet are unknown, zero gradient boundary condition are assumed for all quantities, except the pressure, which is set constant across the outlet. These settings are very simple although not very adequate in this case, because the results show a large recirculation region, which expands through the outlet (see fig. 4). However, to model the flow more accurately it would have been necessary to expand the computation domain further downstream of the draft tube in order to reduce the influence of the unknown outlet boundary condition on the internal flow. This would have increased the computational effort considerably. Therefore the simple boundary conditions have been favoured.

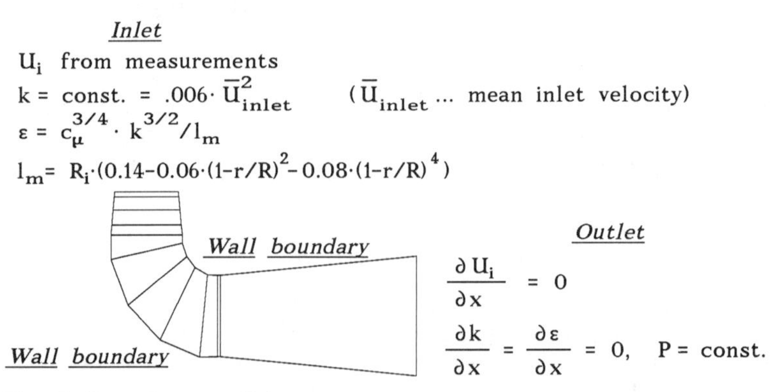

Fig. 3: Boundary conditions

RESULTS

The results of the calculation are presented in fig. 4. The velocity vectors in the mid plane of the draft tube are plotted in fig. 4a. Fig. 4b and fig. 4c show contour plots of the streamwise velocity component at section 11 and at the outlet. It can be seen that a large recirculation region in the diffuser part of the draft tube has been predicted, which expands through the outlet. Therefore, as mentioned above, the outlet boundary conditions have to be considered with caution.

In addition to the outlet boundary conditions, which seem to be the weakest point of the assumptions, the k-ε model also supports the prediction of a large recirculation region. In swirling flows it tends to overpredict the turbulent viscosity causing the swirl to decay too fast. Since in the test rig no measurements could be made, we must rely on comparisons with other numerical solutions to partially ascertain the validity of the present results.

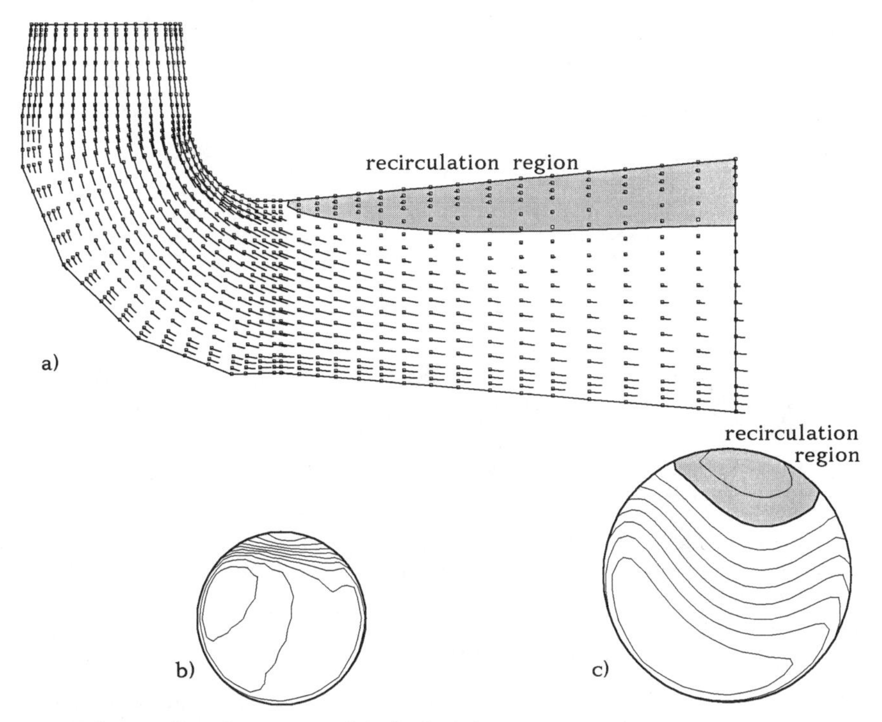

Fig. 4: Velocity distribution in the draft tube,
a) velocity vectors in mid plane,
b) contour of the streamwise velocity component at section 11,
c) contour of the streamwise velocity component at the outlet.

It should be mentioned that no grid independence checks or numerical error estimations have been made, and it is likely that the results are not fully grid independent. The use of a finer mesh, however, has been ruled out for economical reasons.

CONVERGENCE AND COMPUTER RELATED QUANTITIES

The convergence history is shown in fig. 5. There the maximum relative change of the velocity components and the continuity error are plotted versus the number of global iteration steps. The computer related quantities for the calculation are summarized in tab. 2.

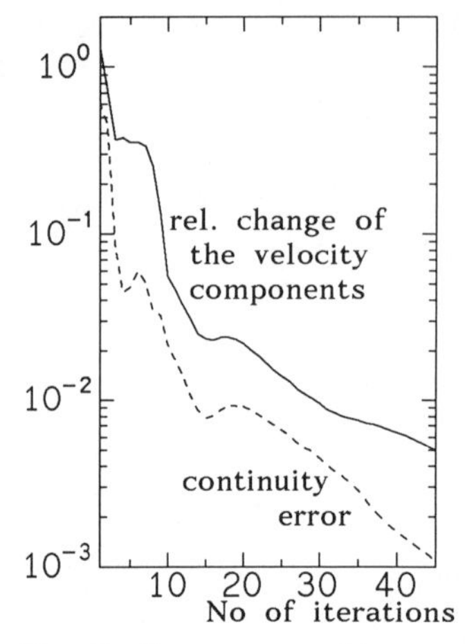

Fig. 5: Convergence history

Tab. 2: Computer related quantities

Computer type	CRAY 2
No of grid points	13772
CPU time	7.5 h
No of iterations	45
CPU time/iteration	600 s
CPU time/node	2 s
CPU time/(node & iteration)	0.044 s
Ratio of vectorisation	--
Parallelisation speedup	--

REFERENCES

[1] BARDINA, J., FERZIGER, J. H., ROGALLO, R. S, "Effect of rotation on isotropic turbulence: computation and modelling", J. Fluid Mech., 154 (1985), pp. 321-336.

[2] NIKURADSE, J.: "Gesetzmäßigkeiten der turbulenten Strömung in glatten Rohren", Forsch. Arb. Ing. Wes, No. 361, 1932.

[3] POURAHMADI. F., HUMPHREY, J. A. C., "Prediction of curved channel flow with an extended k-ε model of turbulence, AIAA J., 21 (1983), pp. 1365-1373.

[4] RODI, W.: "Turbulence models and their applications in hydraulics", IAHR State-of-the-art paper, Delft, 1980.

[5] RODI, W.: "Berechnung turbulenter Strömungsvorgänge in komplexen Kammern", Verbundforschungsvorhaben der VW-Stiftung, Abschlußbericht, 1. Abschnitt, Erlangen, 1988.

[6] RUPRECHT, A.: "Finite Elemente zur Berechnung dreidimensionaler, turbulenter Strömungen in komplexen Geometrien", Dissertation, Universität Stuttgart, 1989.

[7] ZIENKIEWICS, O. C., VILOTTE, J. P., TOYOSHIMA, S., NAKAZAWA, S.: "Iterative method for constrained and mixed approximation. An inexpensive improvement of FEM performance", Comp. Meth. Appl. Mech. Eng., 51 (1985), pp. 3-29.

THREE-DIMENSIONAL TURBULENT FLOW ANALYSIS FOR A HYDRAULIC TURBINE DRAFT TUBE BY THE FINITE ELEMENT METHOD

T. Takagi, S. Tanabe, M. Ikegawa and H. Mukai
Mechanical Engineering Research Laboratory, Hitachi Ltd.,
502, Kandatu-machi, Tsuchiura-shi, Ibaraki, 300, Japan

J. Sato
Hitachi Works, Hitachi Ltd.
3-1-1, Saiwai-chou, Hitachi-shi, Ibaraki, 317, Japan

SUMMARY

This paper analyzes the three dimensional turbulent viscous flow through a given hydraulic turbine draft tube by solving the Reynolds averaged Navier-Stokes equations with a two-equation turbulence model using the finite element technique.

The flow behaviour predicted in the draft tube with a given swirl at the inlet shows that flow separation appears near the top of the draft tube after the bend.

NOMENCLATURE

c_1, c_2, c_d = constants in the turbulence model
k = turbulent kinetic energy
n_i = direction cosine of the normal with respect to x_i axis
p = pressure
U_τ = friction velocity
δU_i = corrective value for a velocity component U_i
$\Delta\tau$ = relaxation parameter
κ = von Karman's constant
ν_t = turbulent kinematic viscosity
Π = pressure ($= p/\rho + 2k/3$)
$\sigma_k, \sigma_\varepsilon$ = Schmidt number for kinetic energy of turbulence and dissipation rate
E = surface roughness parameter
f_i = body force in the x_i direction
n = normal, outward positive
N_i = interpolating functions
r_i = resistance coefficient in the x_i direction
U_i = time averaged velocity component in the x_i direction
$\delta\Pi$ = corrective value for pressure Π
Δt = time increment
ε = dissipation rate of turbulent kinetic energy
ξ_i = local coordinates
ρ = density
ν = kinematic viscosity

INTRODUCTION

Recent progress in computer capabilities has greatly advanced the practical numerical algorithms for computing Navier-Stokes flows bounded by complex geometries [4]. In the case of arbitrarily shaped flow passages the finite element method effectively produces highly accurate solutions for high Reynolds number flows.

A new finite element technique for three-dimensional turbulent flow analysis based on the two-equation turbulence model [1] is applied to the internal flow in a hydraulic turbine draft tube.

The modified ABMAC [3] method is used as the basic algorithm. A sophisticated time integration scheme with second order accuracy and a simultaneous iteration method are proposed to obtain numerically stable and accurate results for both the pressure and the velocity field.

A supercomputer and a parallel computer are used in order to discern the effect of mesh size on the numerical result.

SOLUTION ALGORITHM AND FINITE ELEMENT FORMULATION

The governing equations for three-dimensional incompressible flow analysis based on the two-equation model of turbulence are given in tensor form using summation convention as follows:

$$\dot{U}_i + (U_i U_j)_{,j} = -\Pi_{,i} + \{\nu_{eff}(U_{i,j} + U_{j,i})\}_{,j} + f_i \ , \tag{1}$$

$$U_{i,i} = 0 \ , \tag{2}$$

$$\dot{k} + (U_j k)_{,j} = \{(\nu + \nu_t/\sigma_k) k_{,j}\}_{,j} + \nu_t (U_{i,j} + U_{j,i}) U_{i,j} - \varepsilon \ , \tag{3}$$

$$\dot{\varepsilon} + (U_j \varepsilon)_{,j} = \{(\nu + \nu_t/\sigma_\varepsilon) \varepsilon_{,j}\}_{,j} + c_1 \nu_t (U_{i,j} + U_{j,i}) U_{i,j} \varepsilon/k - c_2 k \varepsilon/\nu_t \tag{4}$$

where

$$\Pi = p/\rho + 2k/3, \quad \nu_{eff} = \nu + \nu_t \tag{5}$$

and

$$\nu_t = c_d k^2/\varepsilon \ . \tag{6}$$

In Eqs (3) and (4) , c_1, c_2, c_d, σ_k and σ_ε are constants, the value of which , used in the Launder-Spalding model [2], are recommended as $c_1 = 1.44$, $c_2 = 0.18$, $c_d = 0.09$, $\sigma_k = 1.0$, $\sigma_\varepsilon = 1.3$.

Table 1 Boundary conditions

inlet boundary	$U_i = \overline{U}_i$; $i = 1,2,3$ $k = \bar{k}$, $\varepsilon = \bar{\varepsilon}$
outlet boundary	$k_{,n} = \varepsilon_{,n} = 0$ $n_i\{\pi - (\nu_t + \nu)(U_{i,j} + U_{j,i})\} = 0$
solid wall boundary	$U_i = 0$ wall function method $U_p = (U_\tau/\kappa)\ln(EU_\tau l_p/\nu)$ where $U_p = \sqrt{U_i U_i}$ $\kappa = U_\tau^2/\sqrt{c_d}$ $\varepsilon = (c_d^{3/4} k^{3/2})/(\kappa l_p)$ n, p, U_p, l_p

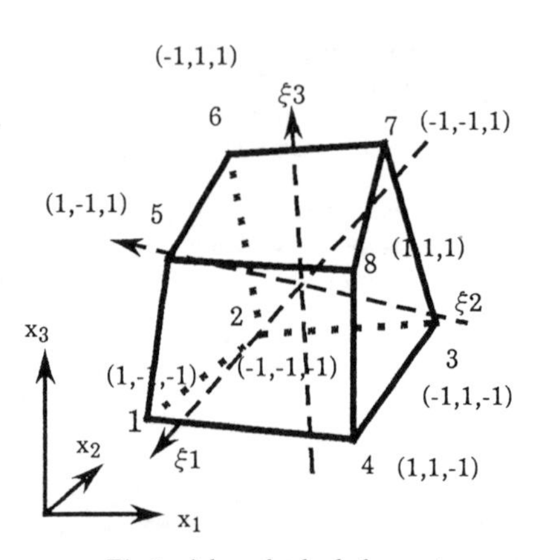

Fig.1 A hexahedral element

The whole boundary S is composed of the inlet boundary S_1, the solid wall boundary S_2 and the outlet boundary S_3. These boundary conditions are given in Table 1.

In order to reduce the number of computing points near the solid wall boundary,the wall function method proposed by Launder and Spalding [2] is employed to provide the boundary condition for k and ε. In Table 1, κ is the von Karman's constant equal to 0.41, E is a function of the wall roughness (9.0 for a smooth wall) and l_p is the length of the normal drawn from the first neighbouring nodal point to the solid wall.

To explain the time integration method briefly, the Navier-Stokes equation (1) is rewritten as

$$\dot{U}_i = F_i - \Pi,_i \tag{7}$$

where

$$F_i = -(U_j U_i),_j + \{\nu_{eff}(U_{i,j} + U_{j,i})\},_j + f_i \quad . \tag{8}$$

Using the Taylor series expansion for U_i, the value of U_i^{n+1} is expressed by

$$U_i^{n+1} \approx U_i^n + \dot{U}_i^n \Delta t + \frac{1}{2} \ddot{U}_i^n \Delta t^2 = U_i^n + (F_i^n - \Pi,_i^n)\Delta t + \frac{1}{2}(\dot{F}_i^n - \dot{\Pi},_i^n)\Delta t^2 \quad . \tag{9}$$

When the Reynolds number is very large, the convection terms are dominant in Eq. (8) and may be approximated by

$$F_i \approx -U_j U_{i,j} \quad . \tag{10}$$

Substituting Eq. (10) into Eq. (9) and making some manipulations, results in

$$U_i^{n+1} \approx U_i^n + (F_i^n + \nu_{jk}^n U_{i,jk}^n)\Delta t - (\Pi,_i^n + \frac{1}{2} \dot{\Pi},_i^n \Delta t)\Delta t \tag{11}$$

where

$$\nu_{jk}^n = -\frac{1}{2} U_j^n U_k^n \Delta t \quad .$$

This momentum equation together with the continuity equation (2) are the fundamental equations. Now an algorithm similar to the ABMAC method for Eqs. (11) and (2) is introduced. First, the intermediate value of U_i^{n+1}, $\tilde{U}_i$ is calculated by

$$\tilde{U}_i = U_i^n + (F_i^n + \nu_{jk}^n U_{i,jk}^n)\Delta t - \Pi,_i^n \Delta t \quad . \tag{12}$$

Next, in order for the values of U_i^{n+1} to satisfy Eq. (2), it is necessary to correct these values as

$$U_i^{n+1} = \tilde{U}_i + \delta U_i, \qquad \Pi^{n+1} = \Pi^n + \delta\Pi \quad . \tag{13}$$

By substituting Eq. (13) into Eq. (11) and subtracting Eq. (12), the corrective values δU_i and $\delta\Pi$ can be obtained as

$$\delta U_i = -\frac{1}{2} \dot{\Pi},_i^n \Delta t^2 \approx -\frac{1}{2}(\Pi,_i^{n+1} - \Pi,_i^n)\Delta t = -\frac{\Delta t}{2} \delta\Pi,_i \quad . \tag{14}$$

By the use of the simultaneous iterations similar to the ABMAC method, the corrective values δU_i and $\delta\Pi$ are calculated as

$$^0U_i^{n+1} = \tilde{U}_i \; , \qquad ^kU_i^{n+1} = {}^{k-1}U_i^{n+1} + {}^k\delta U_i^{n+1} \quad ; \quad k = 1,2,\cdots,L \quad , \tag{15-1,2}$$

$$ {}^{o}\Pi^{n+1} = \Pi^{n} \ , \qquad {}^{k}\Pi^{n+1} = {}^{k-1}\Pi^{n+1} + {}^{k}\delta\Pi^{n+1} \ ; \qquad k=1,2,\cdots,L \ . \qquad (16-1,2) $$

This iteration is continued up to L cycles until the averaged value of the divergence of ${}^{k}U_i^{n+1}$ falls below some small error limit. The relation between δU_i and $\delta\Pi$ in Eq. (14) and ${}^{k}\delta U_i^{n+1}$ and ${}^{k}\delta\Pi^{n+1}$ is given by

$$ \delta U_i = \sum_{k=1}^{L} {}^{k}\delta U_i^{n+1} \ , \qquad \delta\Pi = \sum_{k=1}^{L} {}^{k}\delta\Pi^{n+1} \qquad (17) $$

and the relation between ${}^{k}\delta U_i^{n+1}$ and ${}^{k}\delta\Pi^{n+1}$ is derived from Eq. (14) as

$$ {}^{k}\delta U_i^{n+1} = -\frac{\Delta t}{2} {}^{k}\delta\Pi,_i^{n+1} \ . \qquad (18) $$

By introducing the relaxation parameter $\Delta\tau$ and the following iteration formula for the k+1 iteration values of the pressure and velocity fields, the pressure and the advanced time velocity fields may be solved simultaneously by

$$ {}^{k+1}\delta\Pi^{n+1} = -\Delta\tau\, {}^{k}U_{i,i}^{n+1} \ , \qquad {}^{k+1}\delta U_i^{n+1} = -\frac{\Delta t}{2} {}^{k+1}\delta\Pi,_i^{n+1} \ . \qquad (19-1,2) $$

By repeating these procedures, we can obtain the solution of the Navier-Stokes equations.

The discretized governing equations derived from the three-dimensional flow equations with the two-equation model of turbulence are solved by the finite element method in our analysis. The flow region V is divided into many hexahedral elements as shown in Fig.1. The velocity component U_i and turbulent kinetic energy k and its dissipation rate ε are interpolated linearly within an element by using the values at each node, while the pressure Π is assumed to be constant in each element. The resulting sets of algebraic equations are solved explicitly by the use of the lumping technique for the mass matrix.

NUMERICAL RESULTS

Figure 2 shows the configuration of the given draft tube and two views of the mesh used in the analysis. The mesh was obtained by the body fit method. The number of grid points is 10309 (13×13×61).

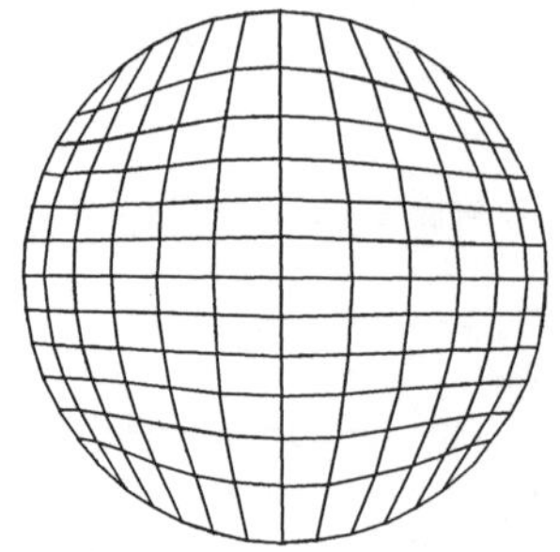

Fig.2(a) Mesh at the inlet section

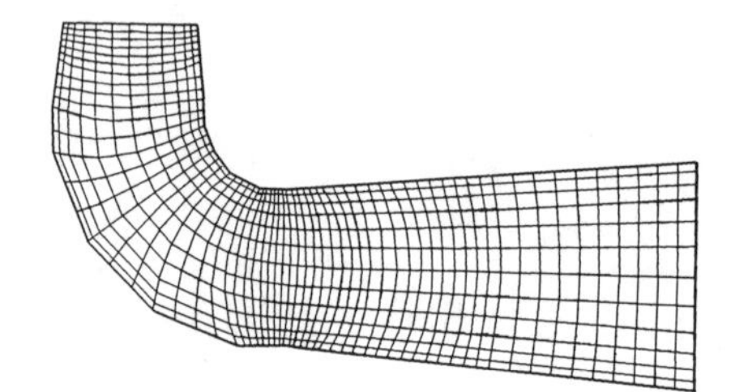

Fig.2(b) Elevation view of the mesh

The numerical results calculated according to the given boundary conditions are shown in Figure 3 and 4. The computer related quantities are shown in Table 2 .

Figure 3 shows the pressure recovery along the draft tube axis. Pressure recovery was very small because of the flow separation near the top of the draft tube after the bend.

Figure 4 shows the given velocity distribution and calculated static pressure distribution along the inlet measurement axis. Static pressure distribution was uniform along the axis at the given conditions of inlet swirl distribution.

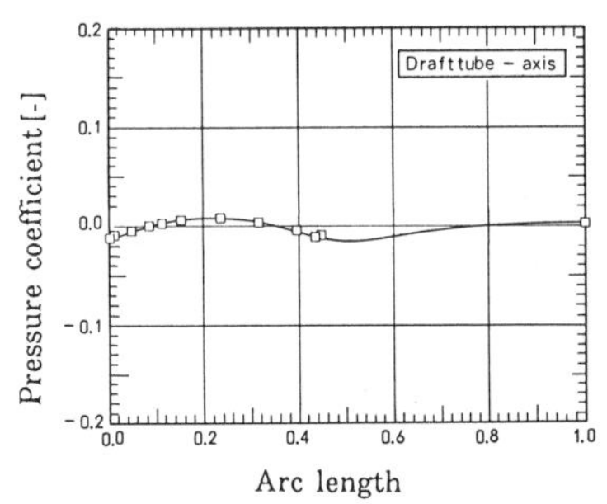

Fig.3 Static pressure distribution along the section's camber line

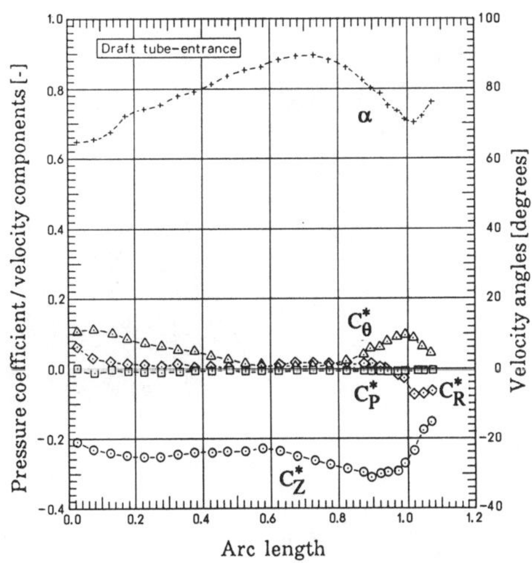

Fig.4 Flow behavior along the inlet measurement axis

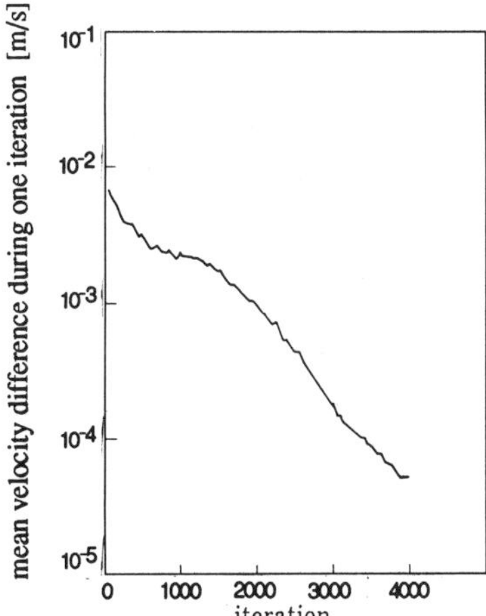

Fig.5 Convergence diagrams

Table 2 Computer related quantities

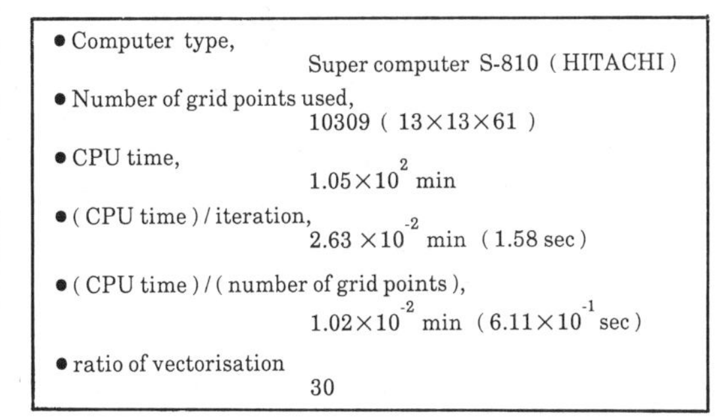

• Computer type,	Super computer S-810 (HITACHI)
• Number of grid points used,	10309 (13×13×61)
• CPU time,	1.05×10^{2} min
• (CPU time) / iteration,	2.63×10^{-2} min (1.58 sec)
• (CPU time) / (number of grid points),	1.02×10^{-2} min (6.11×10^{-1} sec)
• ratio of vectorisation	30

Figure 5 shows the convergence diagrams. In the calculation, the number of iterations was 4000.

Table 3 Computer related quantities

• Computer type,	Parallel computer (CPU;MC68020,68881)×64 64 Processors connected in a two-dimensional array
• Number of grid points used,	45625 (25×25×73)
• CPU time,	1.650×10^{4} min (9.900×10^{5} sec)
• (CPU time) / iteration,	1.650 min (99 sec)
• (CPU time) / (number of grid points),	3.617×10^{-1} min (2.170×10^{1} sec)
• ratio of vectorisation,	No
• parallelisation speedup,	Efficiency ; 99.8%

A new computation has been carried out using a parallel computer as shown in Table 3. The elevation view of the velocity vector field is shown in Figure 6. Very fine meshes with 45625 grid points (25×25×73 nodes) were used. It took 1.65×10^{4} minutes of CPU time for one solution.

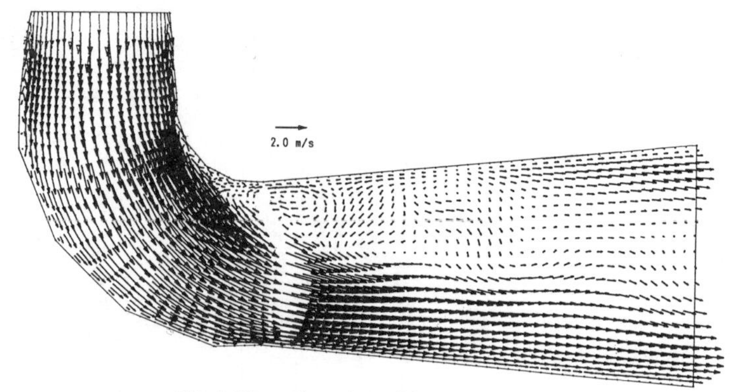

Fig.6 Elevation view of the velocity vector

Comparing to the previous flow field, the flow separation also appears near the top of the draft tube after the bend. However, there are some vortices in the flow field. In the convergence process of the calculation, the vortices were found to move toward the downstream periodically. The reason of this difference is not clear, but it is supposed that the difference is due to the fine mesh employed.

CONCLUDING REMARK

The application of the three-dimensional Navier-Stokes flow analysis with the two-equation turbulence model by the finite element method to hydraulic turbine draft tube was discussed.

This numerical technique proves to be a powerful tool for the design of various kinds of draft tube and complex flow passages.

ACKNOWLEDGEMENTS

The authors gratefully acknowledge the valuable guidance and advice of a senior chief researcher, Dr. A. Uenishi of the MERL, Hitachi Ltd. during the numerical investigation.

REFERENCES

[1] IKEGAWA, M. et. al: " Three-dimensional turbulent flow analysis in a cleanroom by the finite element method", ASME Winter Annual Meeting , FED-Vol.66, 1988, pp. 161~167.

[2] LAUNDER , B.E. and SPALDING , D.B.: "The numerical computation of turbulent flows", Computer methods in applied mechanics and engineering, 3 (1974), pp. 269~289.

[3] VIECELLI, J. A.: "A computing method for incompressible flows bounded by walls", J. of Computational physics, 8 (1971), pp. 119~143 .

[4] VU, T.C. et. al: "Recent development in viscous flow analysis for hydraulic turbine components", IAHR Symposium 1986, vol.2, Montreal, Canada, 1986, pp. 915~926.

3D VISCOUS FLOW ANALYSES FOR THE GAMM WORKSHOP DRAFT TUBE AND FRANCIS RUNNER

T. C. Vu
Dominion Engineering Works, GE Canada
795 First Avenue, Lachine, Québec, Canada H8S 2S8

W. Shyy
University of Florida, Dept. of Aerospace Engineering,
Mechanics and Engineering Science.
231 Aerospace Bldg., Gainesville, Florida, USA 32611

SUMMARY

Three-dimensional turbulent viscous flow analyses for the GAMM workshop elbow draft tube and Francis runner are performed by solving the Reynolds-averaged Navier-Stokes equations closed with a two-equation turbulence model. Implications of both numerical and physical issues on the prediction capabilities have been discussed in the context of design applications.

INTRODUCTION

Since the early 80's, 3-D computation methods based on reduced forms of Navier-Stokes equations have been developed and applied to aid successfully the design optimization of hydraulic turbine components [2,3,9]. For example, full 3-D Euler and potential flow analyses have been demonstrated as excellent design tools for the turbine runner [3,9]. However, those computational methods assuming inviscid flow inside the turbine components, cannot represent the complex behaviour of truly turbulent viscous flows in off-design conditions and cannot predict the associated energy losses.

Recently, intensive efforts have been devoted to developing a suitable numerical algorithm for computing general Navier-Stokes flows bounded by complex geometries [6,1,5,8]. Closed with the $k-\epsilon$ turbulence model ,the Navier-Stokes flow analysis have been applied to successfully predict flow characteristics and energy losses in different non-rotating hydraulic turbine components [10,11,12,13]. It is used regularly by our hydraulic designers to optimize distributor and draft tube geometries. Presently, further application of the viscous flow analysis is being made for hydraulic turbine Francis runners. For the GAMM workshop, the results of 3-D viscous flow analyses performed for the elbow draft tube and the Francis runner are presented.

NUMERICAL ALGORITHM AND BOUNDARY CONDITIONS

The viscous flow analysis is based on the full Reynolds-averaged Navier-Stokes equations. The standard $k - \epsilon$ two equation turbulence model [4] is adopted here as closure form. The numerical formulation comprises a linearized, semi-implicit, conservative finite volume algorithm implemented in a general curvilinear coordinate system. No-slip conditions are applied to all the nodes at solid walls. At the nodal position next to the solid wall, the so-called wall function treatment [4] is used. Strictly speaking, the wall function treatment along the solid surface is valid up to the point of separation. Beyond separation point there is no basis to adopt the wall function treatment anymore. However, in pratice, it has been widely employed even for flows with massive separation and recirculation with the results often, qualitative at least, plausible. Evidences gathered in our previous works for the similar geometrical configurations [10,11,12] indicate that the deficiency of the wall function treatment may not be as severe as they first appear.

At the inlet of the flow domain, the velocity profiles have to be specified. At the exit, zero value of the first order derivatives along the streamwise direction is adopted for all the dependent variables, except for the static pressure which does not require numerical boundary conditions due to the nature of the staggered grid system. This procedure has been studied in depth [7] and found appropriate for the numerical procedure adopted here. As argued in [7], since in the present type of computations the net mass flow rate out of the domain is completely prescribed, it appears that the overall solution is uniquely determined by the known flow rate and the conservation laws, even when the downstream boundaries contain recirculating flow.

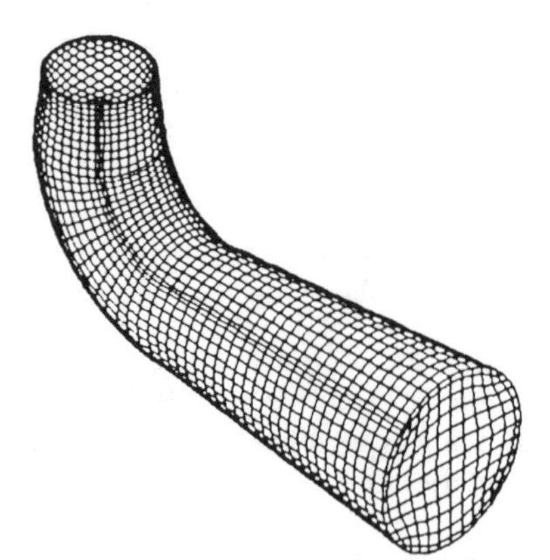

Figure 1. Body-fitted grid system of the smooth elbow geometry.

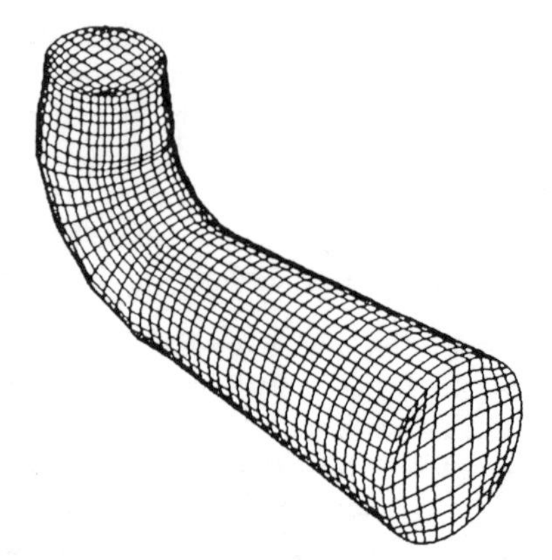

Figure 2. Body-fitted grid system of the original geometry.

VISCOUS FLOW ANALYSIS FOR THE ELBOW DRAFT TUBE

The geometry of the elbow draft tube selected for the workshop consists of an ensemble of welded conical sections. At first, we assumed that the elbow section is fabricated with plastic material as we usually do for our model draft tube. In such a case, the inside surface of the elbow varies continuously as shown in Figure 1 which represents a 3D view of the geometry with the body-fitted grid system. The numerical result presented at the workshop was obtained from the smooth elbow geometry. Since the smooth elbow geometry does not represent correctly the abrupt change at the elbow inside surface, a second flow analysis was performed with a new grid system representing correctely the original geometry as shown in Figure 2. Numerical results obtained from both geometries are presented here in order to demonstrate the strong influence of the elbow surface geometry on the draft tube efficiency and flow behaviour.

The smooth elbow geometry was analysed with three different grid sizes: coarse (15x15x59), medium (19x19x79) and fine grid (25x25x101). The fine grid system represents better the elbow surface variation of the original geometry, therefore its numerical result is presented. The 25x25x101 node grid requires about 69 hours of CPU on a VAX8600 computer. The vectorised version of the code runs 20 times faster on a CRAY X-MP. The convergence criteria is set to 0.01 for the sum of mass or momentum residuals. Figures 3.a and 3.b show the history plots for the mass and momentum residuals and also for the factor of recovery pressure. The original geometry was analysed with a 21x21x87 node grid system.

Figures 4.a to 4.e represent the flow behaviour in the draft tube of smooth elbow geometry. The velocity distribution in the through flow direction as shown in Figure 4.a indicates a large depression of the velocity profile at the outlet but with no flow recirculation. Figure 4.b represents the swirling flow at the inlet section specified by Pitot measurement. Figures 4.c to 4.e represent the evolution of the secondary flow for cross-sections 8, 11 and 12. Numerical results obtained from smaller grid sizes show a smaller depression of the velocity profile at the outlet but simulates qualitatively similar secondary flow characteristics as the fine grid system.

Figures 5.a to 5.e represent the flow behaviour of the draft tube in its original geometry. Observation from the velocity distribution in the through flow direction indicates a large flow recirculation zone starting from the end of the elbow to the draft tube outlet. This is due mainly to the abrupt change of the draft tube ceiling, Figure 5.a, compared to the continuous variation of the ceiling surface of the smooth elbow geometry as shown in Figure 4.a. But the evolution of the secondary flow, Figures 5.c and 5.d, is similar to the previous result. Except at the outlet cross section where the flow recirculation occuring in the top portion, a well defined double swirling pattern is observed.

The most important parameter characterizing the draft tube performance is the pressure recovery factor which is defined as :

$$C_{pr} \quad = \quad \frac{P_2 \quad - \quad P_1}{Massflow\ weighted\ averaged\ kinetic\ energy\ at\ inlet}$$

where P_1 is the massflow-weighted averaged static pressure at the inlet and P_2 is the massflow-weighted averaged static pressure at the outlet. Figures 6 and 7 illustrate

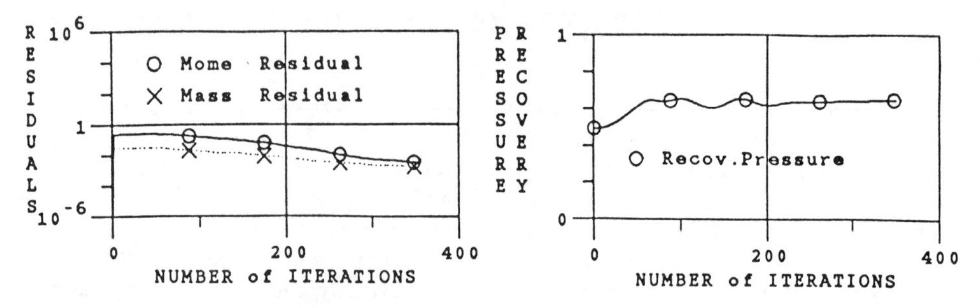

Figure 3.a Convergence history plot of mass and momentum residuals.

Figure 3.b Convergence history plot of pressure recovery factor.

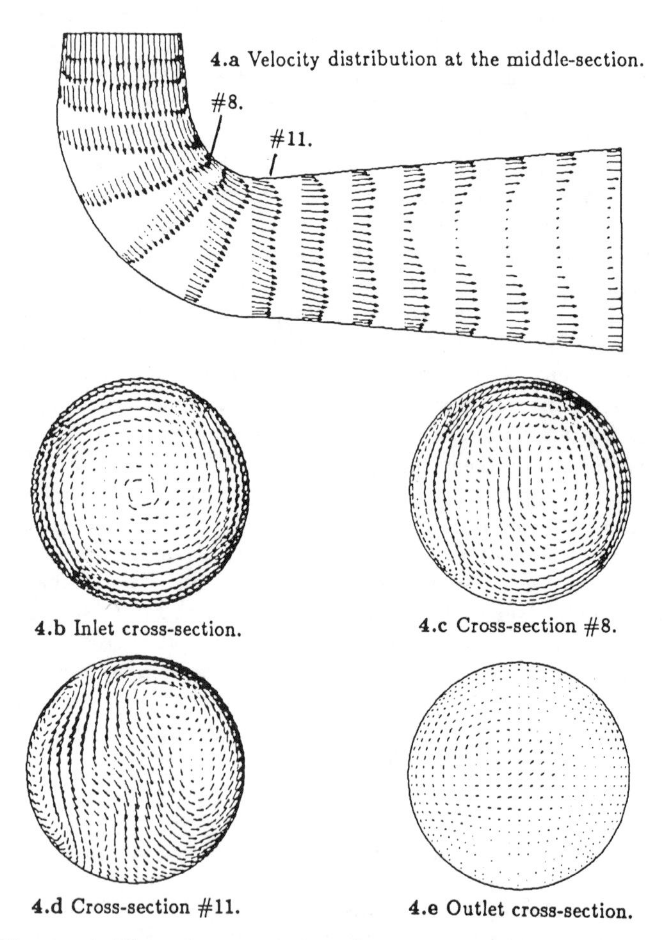

4.a Velocity distribution at the middle-section.

4.b Inlet cross-section.

4.c Cross-section #8.

4.d Cross-section #11.

4.e Outlet cross-section.

Figure 4. Flow characteristics of the smooth elbow geometry.

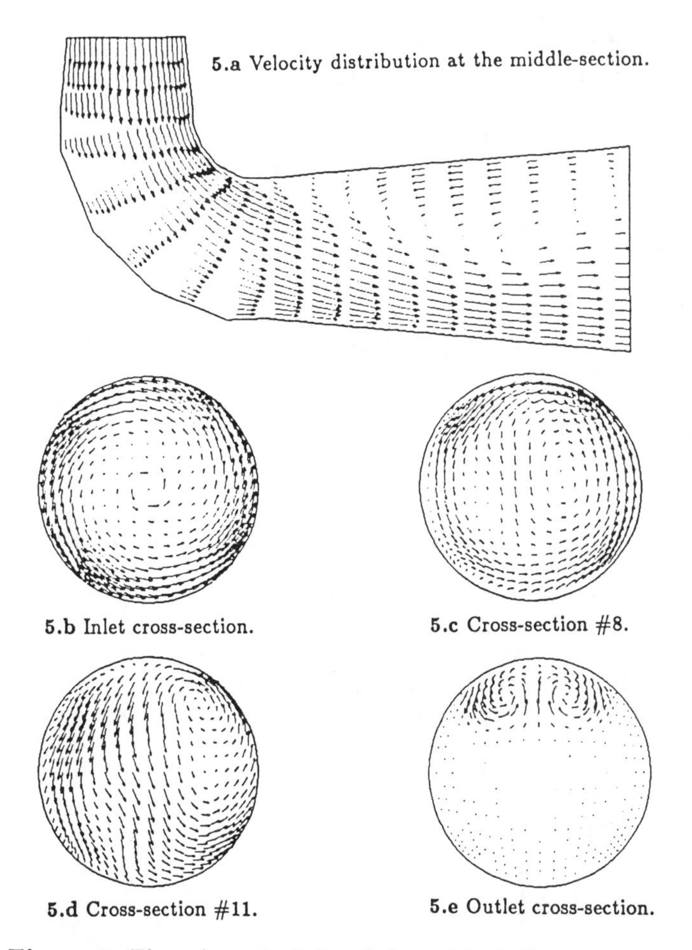

Figure 5. Flow characteristics of the original elbow geometry.

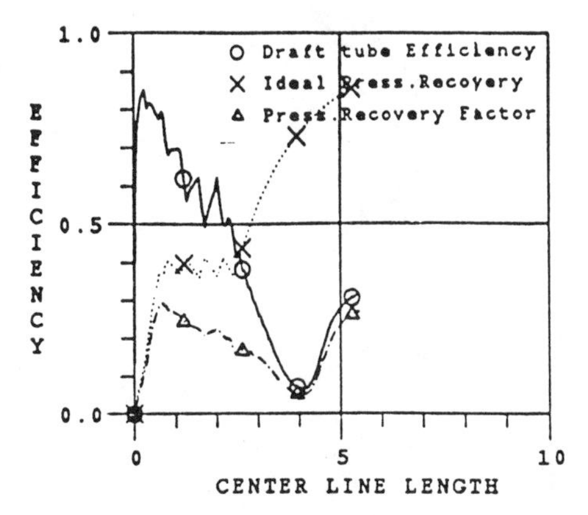

Figure 6. Evolution of the pressure recovery factor - Original geometry.

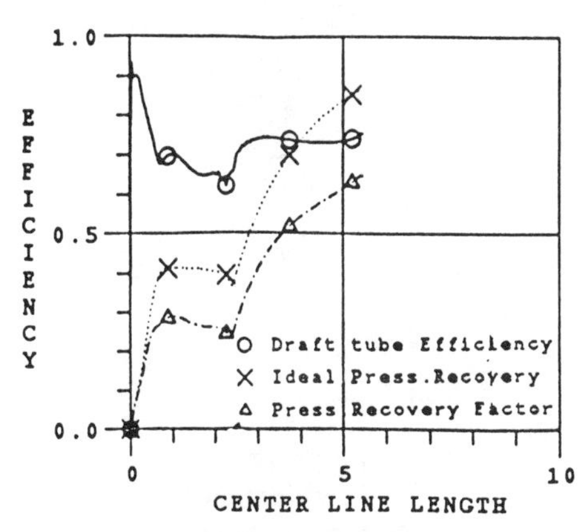

Figure 7. Evolution of the pressure recovery factor - Smooth elbow geometry.

the evolution of the pressure recovery factor from the inlet to the outlet for both geometries. Dotted lines represent the ideal pressure recovery factor based on the cross-sectional area. Dashed lines represent the actual pressure recovery factor obtained from numerical solution. The solid lines represent the efficiency of the draft tube which is the ratio of the actual pressure recovery factor over the theoretical one. Both geometries show the same rates of pressure recovery at the inlet cone region. But due to severe flow recirculation taking place at the end of the elbow region, the pressure recovery factor obtained by the original elbow geometry is about 27% . In comparison, the smooth elbow geometry yields a pressure recovery factor of 65%.

Unfortunately, there is no experimental data on the draft tube flow behavior measurement from the workshop to compare with the numerical results but validation work [11,12,13] performed for the elbow draft tube viscous flow analysis demonstrated that the present numerical model is accurate and reliable for a large range of geometry and operating conditions.

VISCOUS FLOW ANALYSIS FOR THE FRANCIS RUNNER

The application of the viscous flow analysis for hydraulic turbine runners is in the development stage. A rotating coordinate system is adopted and extra forces such as centrifugal and Coriolis forces are added. We present here a laminar flow solution for the Francis runner. The Reynolds number was specified at 800 at which the predicted torque corresponds to the measurement. We are in the process of investigating the effects of various turbulence models on the predicted results.

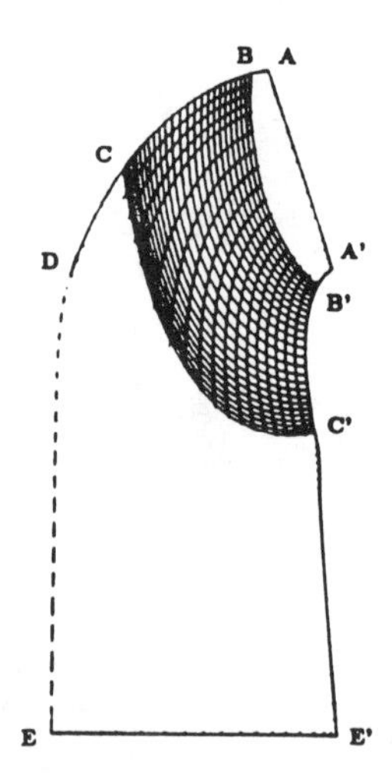

Figure 8. Radial projection of the flow domain with prescribed boundary conditions.

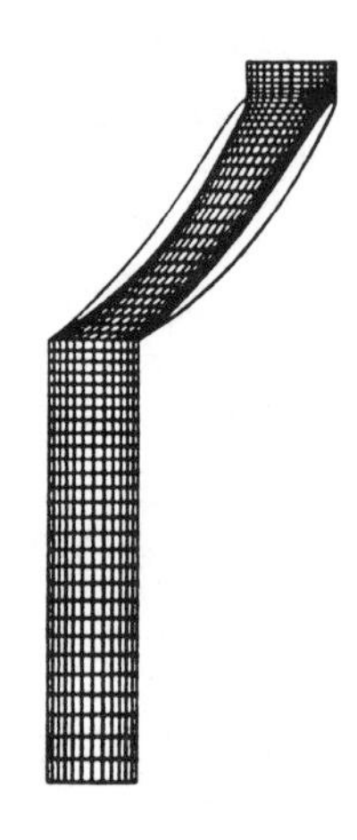

Figure 9. Blade to blade projections of the grid system.

The computational flow domain of the Francis runner consists of a single inter-blade channel including a corresponding portion of the distributor housing at the inlet and a portion of the draft tube inlet cone at the outlet. A radial projection of the flow domain along with prescribed boundary conditions are illustrated in Figure 8. Three types of boundary conditions are applied for the solid walls. In the rotating coordinate system, all the solid walls rotating with the runner, BCD,B'C' and BCC'B', are considered as non-rotating solid surfaces. Non-rotating solid walls such as AB, A'B' and C'E', which represent top and bottom of the distributor housing and the draft tube inlet, are considered as rotating solid surfaces. Surface DE, which is an extension of the hub, is considered as an imaginary solid wall with slip condition prescribed. The periodic boundary condition is applied to the surfaces ABB'A' and CDEE'C'. At the inlet, surface AA', velocity profile has to be specified whereas at the outlet, surface EE', identical extrapolation procedure for velocity, as discussed earlier, is adopted. The pressure does not require any boundary conditions.

Figure 9 shows the projection of the grid system on the blade to blade section. Although being used regularly for the potential flow analysis, this type of grid, which shows an abrupt change in the grid lines at the leading and trailing edges, is not ideal for viscous flow analysis as suggested by the non-smooth pressure distribution. A better body-fitted grid should help correct this problem.

Comparison of the laminar flow solution with experimental data, as shown in the Figure 10, indicates that the predicted pressure coefficient plot at the pressure side agrees with the measurement. But at the suction side, a thick boundary layer prevents the flow from accelerating contrarily to the potential flow solution which always underestimates the pressure coefficient or overestimates the cavitation.

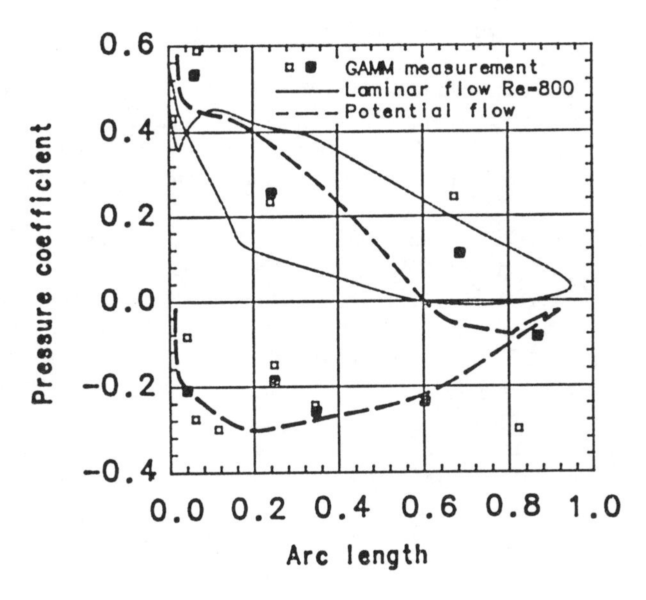

Figure 10. Comparison of pressure coefficient Cp along the section #15.

CONCLUDING REMARKS

The viscous flow analysis, using the $k - \epsilon$ turbulence model, can simulate satisfactorily the flow characteristics and predict accurately head losses in different non-rotating hydraulic turbine components. Development work is underway in order to find a suitable turbulence model to simulate highly turbulent flow in the rotating turbine runners.

REFERENCES

[1] BRAATEN, M.E. and SHYY, W. : "A study of recirculating flow computation using body-fitted coordinates: consistency aspects and skewness", Numerical Heat Transfer, 9(1985), pp. 559-574.

[2] DÉSY, N., DO, H. and NUON, N. : "Turbine casing and distributor design", Water Power 1987, Portland, Oregon, USA, Vol.3, pp.1865-1872, 1987.

[3] DÉSY, N. and DO, H. : "Experience with the application of fully 3D potential flow analysis to runner design", Water Power 1989, Niagara Falls, N.Y., USA, Vol.3, pp.1823-1834, 1989.

[4] LAUNDER, B.E. and SPALDING, D.B. : "The numerical calculation of turbulence flows", Comp. Meth. Appl. Mech. Eng., 3(1974), pp. 269-289.

[5] SHYY, W. : "A study of finite difference approximations to steady state, convection dominated flow problems", J. Comp. Physics, 57(1985), pp. 415-438.

[6] SHYY, W., TONG, S.S. and CORREA, S.M. : "Numerical recirculating flow calculation using a body-fitted coordinate system", Numerical Heat Transfer, 8(1985), pp. 99-113.

[7] SHYY, W. : "Effect of open boundary on incompressible Navier-Stokes flow computation: Numerical experiments", Nume. Heat Transfer", 12(1987), pp.157-178.

[8] SHYY, W. and VU, T. C. : "On the adoption of velocity variable and grid system for fluid flow computation in curvilinear coordinates", to appear in J. Comput. Physics.

[9] THIBAUD, F., DROTZ, A. and SOTTAS, G. : "Validation of an Euler code for hydraulic turbines", AGARD conf. proc. No.437, pp.27/1-27/14, 1988.

[10] VU, T.C. and SHYY, W. : "Navier-Stokes computation of radial inflow turbine distributor", Trans. of the ASME - J. Fluids Eng., 110(1988), pp. 29-32.

[11] VU, T.C. and SHYY, W. : "Viscous flow analysis for hydraulic turbine draft tubes", IAHR Symposium 1988, Trondheim, Norway, pp. 915-926, 1988, also to appear in J. Fluids Eng..

[12] VU, T.C. and SHYY, W. : "Viscous flow analysis as a design tool for hydraulic turbine components", IRCHMB-IAHR Symposium 1989, Beijing, China, pp. 296-307, 1989, also to appear in J. Fluids Eng..

[13] VU, T.C. : "A design parameter study of turbine draft tube by viscous flow analysis", Water Power 1989, Niagara Falls, N.Y., USA, Vol.1, pp.557-566, 1989.

Part 4

SYNTHESIS

CRITICAL EVALUATION AND COMPARISON OF CONTRIBUTED SOLUTIONS

I.L. Ryhming, G. Sottas, A. Bottaro
Swiss Federal Institute of Technology (EPFL)
Hydraulic Machines and Fluid Mechanics Institute (IMHEF)
ME - Ecublens, CH - 1015 Lausanne, Switzerland

INTRODUCTION

The objectives of the 1989 GAMM Workshop, entitled *"3D-Computation of Incompressible Internal Flows"*, were generally defined by the Workshop Scientific Committee to test the ability of researchers in the field to calculate the flow through the different components of a Francis water turbine, including its distributor, runner and draft tube. Consequently, a test case Francis turbine geometry was decided upon and, subsequently, a model turbine was built according to all specifications recommended by the committee for testing in the IMHEF turbine test facilities at EPFL.

With the ambitions clearly defined it was up to the water turbines group, under the direction of Prof. P. Henry, to build an experimental data base including not only integral properties of the turbine flow, e.g. torque, global flow rate, etc., but also detailed measurements of the pressure distribution on the blades of the runner in motion, detailed mapping of the velocity distribution upstream of the entrance to the distributor stay vanes, upstream and downstream of the runner, and, in addition, detailed pressure data for the performance of the draft tube. Unfortunately, not all of these measurements could be carried out successfully. In particular, no experimental data for the performance of the draft tube could be obtained, because highly unsteady flow conditions were encountered, and no experimental equipment was at hand to measure such flows. The unsteadiness of the flow in the draft tube was due to a complicated detachment-attachment phenomenon in the elbow region of the tube itself. With a more careful design of the draft tube this situation could possibly have been avoided. In addition, it was difficult to generate good pressure data on the runner blades. As seen during the Workshop, the data in the most critical region of the runner blades, i.e. following a streamline close to the turbine shroud or "band" on the suction side vary from point to point to such a degree that a comparison with the computed data is all but meaningless. However, and which is very important, the data at the runner outlet are accurate and reproducible, and provide the most useful part of the experimental data base.

Detailed measurements in the entrance region to the turbine (or at the stayring inlet) show non-uniformities in the flow conditions around the circumference at the inlet of the distributor. These non-uniformities are undoubtedly caused by an imperfect design of the spiral casing and cause a series of problems related to, e.g., proper definition of flow conditions at the entrance of the distributor for the calculations. Also, they cause unsteady conditions of the flow in the turbine, which of course influence the pressure distribution on the bladings.

The reasons for all these difficulties were many: this was the first time that such detailed measurements were attempted, and, clearly, the measurement techniques and know-how needed refinement, not realizeable in a short time span. Nor was it possible to redesign the spiral casing or the draft tube, in order to obtain better performances.

Nevertheless, and in spite of these imperfections, some useful lessons can be drawn from the workshop. In the following we shall compare the calculated results with the experimental data and draw some conclusions. In somc cascs comparisons between calculated data only will be possible.

SHORT DESCRIPTION OF THE CONTRIBUTED CALCULATIONS

Each contribution to the workshop is described in detail in the individual papers forming these proceedings. However, for comparison purposes, we list the various contributions in Tables 1, 2 and 3, covering the distributor, the runner and the draft tube calculations, respectively. These tables contain a summary of the methods used in the calculations, as well as some particular features and results.

The test case specifications were designed in such a way that each contributor could choose to calculate separately the flow in the distributor, in the turbine runner or in the draft tube. The turbine flow is assumed to be periodic in the circumferential direction, so that only one interblade flow passage needs to be investigated. All contributors have made the same assumption for the distributor as well as for the runner. Indeed, this was also intended to be the case in setting up the test case, but was, unfortunately, not realized in practice.

The possibility was also offered to do a simultaneous calculation of the flow in a combination of two parts by using a suitable method (e.g. multiblock), or to do the entire turbine. Nobody furnished a simultaneous calculation of the complete Francis turbine configuration but two contributors solved, by using a multidomain strategy, for the flow in the distributor and the runner.

The short description of the various contributions that follows clarifies the content of the Tables.

DISTRIBUTOR AND RUNNER SIMULTANEOUS CALCULATIONS

Two authors, Eliasson and Goede, contributed a simultaneous simulation of the flow in the distributor and the runner. Their contributions are summarized in Table 1 (distributor part) and 2 (runner part). They adopted some kind of multiblock technique in conjunction with an explicit finite volume scheme to integrate the Euler equations of motion. The integration of the continuity equation is based on Chorin's artificial compressibility concept, i.e. a steady state solution is found by means of a (pseudo-)time marching procedure. This technique has become a rather popular way to obtain a steady incompressible solution of the Euler equations.

In Eliasson's contribution, the distributor is divided into two blocks, and an additional block covers the runner. He uses the measured velocity distribution at the entrance to the distributor as inlet boundary condition to his three-block computational domain. At the exit of the runner characteristic boundary conditions are prescribed. Interface conditions are used to transfer information from one block to the other. Hence, in formulating the conditions for the runner block, which is in relative movement with respect to the wicket gate block, an averaging procedure, specifying *no tangential variation* of the flow velocity, is assumed.

In Goede's contribution, the approach is based on the same principles as described above. However, his technique of "stacking" the three domains (two for the distributor and one for the runner) is different from Eliasson's multiblock concept. First of all, Goede assumes that the entrance velocity to the distributor is constant, which would be the case for an ideal spiral casing, and fixes the outlet pressure at the exit of the runner to be constant (P_{ref}). Then, he uses the pressure computed at the inlet to the runner as the downstream boundary condition for the wicket gate "stack", and so on. Overlapping of the "stacks" (or grids) is said to be "beneficial" in regions of strong interaction between the "stacks", e.g. at the entrance to the runner. Hence, in this region he extends the wicked gate grid downstream, and the runner grid upstream to obtain an overlap region between these two "stacks".

Distributor calculations

In Table 1 the four contributing groups with results for the flow in the distributor are listed. Eliasson's and Goede's works have already been presented in the previous subsection. Thus, we restrict ourselves here to the two contributions relating to the flow simulation in the distributor only.

In the Bottaro et al.'s contribution, one single block was adopted for the distributor calculation and the same procedure for the integration of the Euler equations was used as in Eliasson's and Goede's contributions. Measured inlet velocities along the vertical axis probed in the experiments, plus an assumed axisymmetric velocity distribution, were used as inlet Dirichlet boundary condition. At the exit of the distributor the experimentally determined pressure distribution was imposed.

In their contribution, Nagafuji et al. solve the Navier-Stokes equations by a finite volume technique in which the pressure-velocity coupling is treated by using the SIMPLE algorithm. The viscosity used in the calculation is taken to be 100 times larger than its real physical value. Measured data are used as boundary conditions over the inlet and exit sections.

Runner calculations

Table 2 contains the 13 contributions that solved for the flow in the runner. Evidently, the runner flow prediction drew the greatest interest at the Workshop. Of these 13 contributions, 3 are based on the potential flow assumption, 9 are based on the Euler equations and one contribution is based on the Navier-Stokes equations.

The potential flow contributions were based on either a finite element treatment or a mixed finite volume-finite difference formulation.

The Euler contributions contain three principal variants, namely those using the explicit finite volume technique described briefly in the previous subsections, those using the Clebsch formulation in the finite element approach, and one contribution which adopts an implicit finite difference method.

In the single contribution for which the Navier-Stokes equations are solved, a semi-implicit finite volume technique is adopted to solve for the *laminar flow* in the runner with a Reynolds number equal to 800.

In addition to the content of Table 2, it should be mentioned that the contributions of Eliasson, Goede and Bottaro et al. follow the lines described earlier.

The contribution of Billdal et al. also follows the same lines but distinguishes itself by the treatment of the boundary conditions at the entrance of the runner, where, by using *characteristic variables* rather than *physical* ones, the measured global flow rate can be satisfied as well as the proper distribution of the total pressure.

This is in contrast to the Bottaro et al. contribution, where it is remarked that measured global flow rate and calculated flow rate obtained by using measured profiles and the assumption of axisymmetry differ by 7%. Hence, two calculations have been performed by Bottaro et al.; in one the measured velocity profiles are used as inlet conditions, and in the other one "adjusted" inflow data are prescribed to match the measured flow rate. In this second case the "adjustment" is simply made by rescaling the absolute velocity by a constant factor; by so doing obviously the relative flow angle β is modified. Another adjustment methodology, valid for Euler calculations, is proposed by Kubota (see Part 2); it consists in normalizing the measured data to keep the conservation of discharge, angular momentum and total specific hydraulic energy after having suitably eliminated all boundary layers.

Grimbert et al. use a Clebsch formulation leading to a finite element discretization of the resulting basic equations. The implementation of the boundary conditions becomes again a major problem, because of differences in measured global flow rate and calculated ones, based on the integration of the experimental normal velocities at inlet and outlet. The approach adopted by Bottaro et al. to handle this problem is essentially taken over by Grimbert et al. Two sets of results for the runner are presented.

In the approach used by Liess and Ecer the finite element code PASSAGE, based on the Clebsch variables, is applied to the runner flow. Measured data are used to formulate the required boundary conditions in this approach. The PASSAGE algorithm is based on a block structuring technique rather than on the assembling of the equations on the entire computational domain.

Arakawa et al. use a numerical algorithm inspired by Beam and Warming to advance the hyperbolic system of equations, in the artificial compressibility formulation, in time. The spatial differencing of the explicit part is done by using an upwind TVD scheme. The authors utilize this scheme, because it minimizes the artificial dissipation errors even for an incompressible flow.

The potential flow approach is adopted by Nagafuji et al. to calculate the flow in the runner. The numerical scheme is based on a finite element discretization. The normal velocities derived from the experimental data and corrected to satisfy the specified mass flow are used as inlet and outlet boundary conditions. Two sets of results are produced on the assumption of "free vortex flow" and "quasi-free vortex flow". The distinction between these two concepts is not clear. However, in applying the "free vortex flow" concept, the circulation in the vaneless spaces at inlet and outlet of the runner is taken to have constant but different values.

In the potential flow approach used by Lymberopoulos et al., centered finite-difference/finite-volume schemes are applied to discretize the basic equations. All Neumann type boundary conditions on the potential function are treated implicitly via a flux-balance procedure.

Finally, the Navier-Stokes equations are considered by Vu and Shyy to solve for the laminar flow through the runner. In this approach a semi-implicit finite-volume scheme is used, and the Reynolds number is chosen such that the predicted torque corresponds to the measured value.

DRAFT TUBE CALCULATIONS

Seven contributions to the Workshop were concerned with the flow in the draft tube. Such calculations are listed in Table 3. All except one of these contributions are based on the Reynolds averaged Navier-Stokes equations, and the eddy viscosity concept is used. The last contribution is based on the Euler equations.

Kubota et al. solve the equations by using a finite element technique in which 10-node tetrahedral elements are employed. The eddy viscosity is approximated by two separate and distinct values, one in the core region and the other near the walls.

Lazzaro and Riva use the FIDAP code employing 8-node brick elements. The eddy viscosity is calculated using the standard k-ε formulation. The wall function approach is a built-in feature of the code, and default boundary conditions (specifying vanishing streamwise gradients) are applied over the exit section.

Ruprecht contributes two calculations employing different pressure conditions at the diffuser outlet. He uses the FENFLOSS code, which employs 8-node brick elements. The standard k-ε formulation with the wall function approach provides the eddy viscosity.

Takagi et al. use the ABMAC finite element method, in which hexahedral elements are employed. Again the standard k-ε formulation with the wall function is adopted to obtain the eddy viscosity.

The Vu and Shyy approach is the same as the one they used to calculate the runner flow. However, for the draft tube analysis they solved for the governing equations with the k-ε turbulence model.

All these Navier-Stokes contributions are therefore similar being based on the same theoretical basis. The differences lie on the boundary conditions applied at the draft tube exit, as well as on the use of different elements and mesh structures, see Table 3. Hence, a maximum of 63125 elements is used by Vu and Shyy in one of their calculations, and a minimum of 6948 elements is used by Lazzaro and Riva.

Finally, Liess and Ecer attempt the use of an Euler code to calculate the draft tube flow, being well aware of its limitations in a case where only a fully viscous analysis is adequate. They consider this analysis as a first step towards a more consistent approach to the problem. Not unexpectedly, they run into convergence difficulties with their Euler calculation when the mesh density is increased beyond a certain value. Their algorithm makes uses of Chorin's artificial compressibility concept for the pressure-velocity coupling and of an explicit finite volume procedure to integrate the resulting system of equations.

DISTRIBUTOR CALCULATIONS : COMPARISON WITH EXPERIMENTAL DATA

The three Euler calculations for the flow in the distributor differ in the implementation of the boundary conditions and in the definition of the geometry at the trailing edge of the vanes' blades. Because of the numerical technique adopted by these three groups, inlet velocities are specified, while inlet pressures are computed. However, because of the difference between single block and multi-block calculations the pressure at the exit of the distributor is either imposed as boundary condition (as in Bottaro et al.) or computed (as in Eliasson and Goede). Note also that all contributors have used H-H grids with nearly the same density.

At the inlet Goede ignores the detailed distributions that have been measured along a single vertical axis, and assumes constant flow conditions for c_r, c_θ, c_z and α, see Fig. 1. Eliasson and Bottaro et al., on the other hand, matches the inlet flow data very well (see Fig. 1). The inlet pressure is computed; it is slightly underestimated by Goede and overestimated by an average of 10% by Bottaro et al. Eliasson finds a very good agreement with the measurements, Fig. 1.

At the outlet, Eliassons' predictions are very good : c_p is slightly underestimated, c_θ is overlapped to the experimental data, α experiences a parallel shift of about 2°, c_z and c_r are very slightly below the measurements (Fig. 2).

The Bottaro et al. outlet results are such that c_θ is slightly overestimated, c_r is weakly underestimated all along except for $s \geq 0.55$, where it becomes greater than what experimental data indicate; the predicted c_z deviates even more from the data, and the α prediction is good except for very small and very large α's.

Goede's outlet prediction for c_p is very good, c_θ is in reasonably good agreement and much better than the estimate for c_z and c_r. The flow angle α is in fair agreement, underestimated for the smaller arclengths and overestimated for the larger ones.

It is interesting to see that these differences in results depend on a different interpretation and handling of boundary data. It should be kept in mind that both Goede and Eliasson use variants of a multiblock approach. Hence, the upstream influence of the runner is detected in the distributor calculation. Subsequently, we shall see the effects of this approach in comparison with the single block approaches adopted by the majority of the contributors using the Euler equations, when examining the runner results. It is clear that the adoption of Dirichlet conditions at the entrance of a single block (the runner for instance) and, in particular, in close proximity to the leading edge of the blades, produces waves that reflect back toward the inlet boundary and are not absorbed by the boundary itself. This procedure can be the cause of significant errors.

The Nagafuji et al. approach is based on the Navier-Stokes equations. As boundary conditions, they impose the outlet experimental pressure distribution and the inlet experimental velocity distribution, but with constant c_z. At the inlet the computed c_p's match the measured values very precisely (Fig. 3). At the outlet their predictions are such that c_θ is underestimated by 15% or more, while c_z and c_r are in somewhat better agreement with the experiments. The angle α is predicted well only until $s = 0.25$, afterwards large deviations follow as evidenced in Fig. 4. Their analysis, as they admit, represents a crude way to take into account viscous effects, since the viscosity (molecular and eddy) is simply replaced by a constant (and large) value. Hence, they compute a flow with $Re = 1000$ only (based on inlet height of the distributor and average speed over the inlet). In short, these predictions are more expensive and more or less comparable to the Euler results. It is nonetheless encouraging that a viscous approach has been attempted with a fair degree of success. This should lead the way to more complex turbulent analyses.

As a final comparison, in Fig. 5 we show the results of all four contributing groups for the pressure distribution at midheight through the distributor (Z=0), along pressure and suction sides of stay vane and wicket gate. The results are comparable and are representative of all the results obtained by the various groups at different levels Z. It seems, however, that Bottaro et al. consistently overestimate c_p. Also, it appears that the pressure near trailing edges shoots up or down because of the different treatements of the geometry there[1]. None of the Euler contributors enforced explicitely the Kutta condition at the blades' trailing edges.

RUNNER CALCULATIONS : COMPARISON WITH MEASUREMENTS

For easy reference in comparing the results, the contributed solutions have been numbered as shown in Table 2. To begin with and for comparison purposes, we choose to regroup together solutions which might have some common features. Hence, a first group is formed by contributions {2}, {5} and {6}; a second group is formed by contributions {3}, {7} and {11}; a third group is formed by contributions {4}, {8} and {12}, and a fourth group is made up of contributions {1}, {9}, {10} and {13}.

In the following, the comparison of the computed results in each group will be shown in a series of figures containing *a)* the c_p distributions along the three theoretical streamline sections specified on the blades, and *b)* the velocity components, flow angles, and pressure distributions along the probed inlet, outlet and middle axes.

The comparison of the results of the first group are shown in Figs. 6-9. In Fig. 6 the measured pressure data along the three sections 2, 9 and 15, are compared with computed results. It is immediately evident that the differences between the calculated results is

---

[1] We should recall that in the experimental geometry the blades' trailing edges are not sharp but rectangular.

relatively small. In section 2, the difference between calculated and measured data is well noticeable on the suction side for $s > 0.2$. This feature is apparent as well in sections 9 and 15. It is difficult to find an explanation for this rather consistent difference, since the computed torque values agree well with the measured one as shown in Table 2.

Some differences in the calculated distributions are, nevertheless, clearly distinguishable. Goede {6} found a strong suction peak on the low pressure side of the blades along section 15. Indeed, in the experiments a small cavitating zone was observed in this area of the blades, Goedes' $c_{p\ min} \approx -0.3$ agrees favourably with this observation. The other calculations in Fig. 6 fail to reveal such a phenomenon.

As a result of these pressure calculations, it is interesting to compare the measured with the calculated torque, the latter being obtained from the calculated pressure distribution. The best result in this context is obtained by Eliasson {5} and it differs from the experiments by 4.30%.

Next, in Figs. 7-9 the results at the inlet, middle and outlet axes are compared. We recall here, that in {2}, characteristic variables have been used in specifying the boundary conditions at the inlet, whereas in {5} and {6}, different variants of multiblock techniques have been used to transfer data between the runner block and the two blocks upstream constituting the distributor. At the exit, characteristic variables have been used in {2} and {5} and a constant pressure value has been assumed in {6}.

It is interesting to see the effect the various numerical approaches have on the results. At the inlet, the velocity data for c_a, c_u and c_m, as well as the velocity angles and the c_p distribution are in excellent agreement in all three contributions, with, perhaps, the best results obtained in {2}. This tendency is maintained at the middle axis as well as at the outlet. The good quality of the results of Billdal et al. {2} should be ascribed at least partially to the *characteristic variables* treatment of the inflow boundary conditions[2]. The results of {5} and {6} are also very good because the upstream influence of the distributor is maintained through the multiblock technique, even if, in the case of Goede, a simplified irrotational velocity distribution was assumed at the entrance of the distributor.

The results obtained in the second group are shown in Figs. 10-13. As can be seen in Fig. 10 the spread in the calculated pressure data in the streamline sections 2, 9 and 15 is considerably greater than in the previous group. The least successful calculation in reproducing the measured data seems to be that of {7}.

The potential flow results of {11} capture remarkably well the suction peak in section 15, but the subsequent pressure recovery is too high and a "fish tail" like pressure distribution is exhibited towards the trailing edge.

The differences in the calculated results are considerable, which is also clearly visible in the calculated torques showing a spread of 13.25%. The best result in this context is obtained in {3}, where the calculated and measured torque differs by -5.11%.

---

[2] A good design of the computational mesh is certainly of equally great importance, but we reason here on the assumption that all the grids employed by the various contributors were equally well constructed.

In Figs. 11-13 velocities, velocity angles, and pressure distribution at the inlet, middle and outlet axes are compared. It is useful here to recall the differences in applying the boundary and initial conditions in the various calculations. At the inlet the measured velocity distribution is used in {3} and {7}. This seems to be the case also in {11}, where, by carefully choosing the inlet circulation and imposing the normal velocities, nearly all details of the measured distributions can be captured except in a region close to the upper wall. There are also differences in the c_a distribution, as can be seen in Fig. 11. The measured inlet pressure is imposed as boundary condition in {7}, whereas {3} and {11} compute the inlet pressure and important deviations from measurements are found in both cases.

With these differences in mind it is interesting to note how successful the potential approach of {11} is in reproducing the measurements in the middle and outlet axes positions (Figs. 12 and 13). Certainly, the details in the distributions close to the lower wall are not caught, but in general the potential results are better than those obtained with these particular Euler solutions, where the boundary conditions have been applied in a different way in comparison with the previous set of solutions. At the outlet the importance of a "proper" outflow boundary condition is stressed by the poor pressure predictions of {7}. Contributions {3} and {11} impose the measured pressure as Dirichlet boundary condition.

In the following series of Figs. 14-17, the results of the third group, i.e. {4}, {8} and {12}, are compared. This series differs from the previous one because of the type of boundary conditions that have been applied at the runner entrance.

As can be seen the results are, in general, of better quality. Already, the differences between the calculated pressure distributions are much less pronounced, Fig. 14. Only the suction side calculations in section 15 differ to some extent. Surprisingly enough, the Euler solution of {4} and the potential solution of {12} agree better with one another than the Euler solutions of {8} and {4}. In comparison with the measured data, it is seen that the predictions on the suction side do not agree as well as those on the pressure side, a feature which is confirmed by all contributors to this test case.

The computed torque values are surprising : in {4} the difference with respect to the measured value is +15.92%, in {8} +13.73% and in {12} +14.83%. Since the computed pressure results agree better with the measured data in the latter group than in the preceding one, some doubts are cast on either the quality of the pressure measurements or on the validity of the computed torque values.

Bottaro et al. and Grimbert et al. modified the c_a velocity distribution at the runner inlet to match correctly measured and numerical flow rates. Due to these changes the velocity components and the velocity angles do not correspond with the measurements as well as in the previous series of calculations. However, better results for the pressure distribution are obtained in general, see Fig. 15. Along the middle and outlet axes the Euler results of {4} are in good agreement with the measurements, the Euler results of {8} do not match as well, and the potential flow results of {12} are partly of good quality.

The results of the last group are shown in Figs. 18-21. The computed c_p distributions differ from one another to a considerable extent, see Fig. 18. However, the two Euler solutions {1} and {9} seem to reproduce the data better than the potential solution {10} and the Navier-Stokes solution {13}. In terms of the torque, {1} is off by -2.41%, {9} is off by

+2.83%, {10} is off by -21.75%, and {13} is off by only -0.49%. Indeed, it is difficult to understand the latter result in examining the calculated pressure distributions.

In Figs. 19-21 the results along the inlet, middle and outlet axes are shown. Along the inlet axis, Fig. 19, the measured velocity distribution is used in {1} and {13}. In {13} the presence of a boundary layer is noticeable at the bounding walls. Less accurate matching of the components of the velocities are obtained in {9} and {10}. The c_p are in good agreement with the measurements except for {13}.

Along the middle and outlet axes, shown in Figs. 20 and 21, only {1} and {9} agree reasonably well with the measured data. The other two solutions deviates to a considerable degree from the measurements. The best results obtained in this group is {1} in predicting the flow downstream of the runner.

The Navier-Stokes solution shows a large overprediction of the c_u velocity component close to the band, although $c_u = 0$ at the walls. The other two components are oscillating close to $s = 0$. The problem is perhaps, that a sufficiently fine mesh was not employed, in addition to the inadequacy of the "turbulence approach" taken. Moreover, this solution exhibits poor c_p predictions.

DRAFT TUBE CALCULATIONS : COMPARISON OF PREDICTIONS

The experimental observations of the draft tube flow indicate that large pulsations occur at the exit, and these were probably caused by a complicated flow separation-reattachment phenomenon at the elbow upper bend. Hence, the flow in the draft tube was certainly unsteady in character, a fact that none of the contributors have taken into account in performing the calculations or, indeed, have guessed as a result of analysing the results. Nor did anyone attempt an unsteady flow computation.

The flow fields that have been calculated differ considerably from one prediction to the other, but the majority of contributors has found a *separated flow* with a large recirculating region extending from the upper elbow bend all the way to the exit. In this recirculating region Takagi et al. have found *transverse* vortices downstream of the separation point. In the convergence process of the calculation, they even detected that the vortices moved towards the exit periodically on the finest mesh that was used, a possible indication of unsteadiness, even if the calculation converged. Another solution where vortices were found is that presented by Vu and Shyy, where two counterrotating *longitudinal* vortices are present at the exit section.

An important point raised by several of the contributors is the prescription of the *downstream boundary conditions* that would allow for recirculating flow. In general the pressure was set to a constant, and for the other dependent variables a variety of conditions are proposed, see Table 3.

All contributors that used the k-ε model to calculate the eddy viscosity employed the wall function approach. This is certainly questionable in regions where the flow is strongly three-dimensional, i.e. close to the walls and in recirculating regions involving a high degree of swirl. Some of the contributors remarked this fact, but no other turbulence model was adopted and compared.

The comparative results on the draft tube are shown in Figs. 22-25. A fair degree of agreement if found for the c_p along the draft tube axis. Nonetheless, pictures of the secondary flows at different downstream stations presented in individual contributions differ considerably.

FINAL REMARKS

The flow in a Francis turbine configuration has been calculated by several techniques. Thc turbinc chosen is characterized by low efficiency (at the chosen operating point) and a bad cavitation behaviour. However, the velocity distribution at the exit of the runner exhibits good characteristics, such as an almost constant meridional component and a peripheral velocity component slightly increasing towards the wall.

Most of the calculations failed to pinpoint a low pressure peak near the band on the blade suction sides, where cavitation in the runner is experimentally found. The notable exceptions are the Euler contribution by Goede and the potential flow contribution by Nagafuji et al. This fact points out two essential things that a computational fluid dynamicist must deal with : *boundary conditions* and *grids*.

In the case of the runner we should try to avoid the specification of inflow boundary conditions too close to the leading edge of the blades to prevent reflexion of waves. Also, outflow boundary conditions for incompressible flows remain an open problem.

Concerning grids, it is noteworthy that identical solvers for the flow simulation yielded rather different solutions. This indicates that a proper grid, with qualities such as orthogonality, limited stretching of control volumes, etc... (for an inviscid solver), might in effect make the difference between acceptable and unrealistic results.

A global overlook at the results contributed reveals that :

- Euler equations-based solvers are adequate to predict accurately the flow in the components of a Francis turbine (except for the draft tube);
- best results for the flow in the runner are obtained by multiblock techniques or by methods which do not fix the inlet flow via Dirichlet conditions (which *perfectly reflect* waves), but let them evolve with the solution procedure;
- in view of the fact that the inviscid Euler results are remarkably good it is doubtful whether viscous analyses of distributor and runner could provide any additional insight into the flow field;
- viscous analysis are indispensable for the prediction of the flow in the draft tube and renewed interest is being spent on turbulence modelling to take into account effects of swirl, separated flow and unsteady mean flow.

Table 1. Distributor calculations - Brief overview of the contributions.

CONTRIBUTOR(S)	BASIS	SCHEME	DOMAIN DECOMPOSITION	MESH SIZE[1]
Bottaro / Drotz / Gamba / Sottas / Neury	Euler	Explicit F.V., artificial compressibility	Single block	$105 \times 17 \times 17$ [2]
Eliasson	Euler	Explicit F.V., artificial compressibility	2 blocks for the distributor + 1 block for the runner	$2 \times (49 \times 25 \times 13)$ [2]
Goede	Euler	Explicit F.V., artificial compressibility	2 "stacks" for the distributor + 1 "stack" for the runner	$2 \times (42 \times 20 \times 12)$
Nagafuji / Suzuki / Kobayashi / Taniguchi	N.S. $\nu = 100 \cdot \nu_{real}$	Implicit F.V., SIMPLE	Single block	$75 \times 30 \times 30$

[1] All the contributors have used H-H mesh topology.

[2] The effect of mesh size was investigated; the mesh indicated is the finest used.

Table 2. Runner calculations - Brief overview of the contributions.

	CONTRIBUTOR(S)	BASIS	SCHEME	DOMAIN DECOMPOSITION	MESH SIZE[1]	TORQUE[2]	REMARKS
{1}	Arakawa / Samejima / Kubota / Suzuki	Euler	Impl. F.D., artificial compressibility	Single block	$65 \times 21 \times 21$	0.24960 (-2.41%)	
{2}	Billdal / Jacobsen / Bratsberg / Andersson / Brekke	Euler	Expl. F.V., artificial compressibility	Single block	$78 \times 15 \times 20$	0.27753 (+8.51%)	Global continuity satisfied by using characteristic variables at inlet
{3}	Bottaro / Drotz / Gamba / Sottas / Neury (1)	Euler	Expl. F.V., artificial compressibility	Single block	$65 \times 17 \times 21$[3]	0.24270 (-5.11%)	Inlet b.c. using the provided data
{4}	Bottaro / Drotz / Gamba / Sottas / Neury (2)	Euler	Expl. F.V., artificial compressibility	Single block	$65 \times 17 \times 21$[3]	0.29650 (+15.92%)	Inlet b.c. using the provided data and adjusting the flowrate
{5}	Eliasson	Euler	Expl. F.V., artificial compressibility	2 blocks for the distributor + 1 block for the runner	$113 \times 25 \times 21$[3]	0.26678 (+4.30%)	
{6}	Goede	Euler	Expl. F.V., artificial compressibility	2 "stacks" for the distributor + 1 "stack" for the runner	$42 \times 12 \times 20$	0.26903 (+5.18%)	Constant flow field at inlet stay vane ring
{7}	Grimbert / Verry / El Ghazzani (1)	Euler	Clebsch, F.E. (Euler 3D)	Single block	1862 20-noded quadratic elem.	0.23831 (-6.83%)	Inlet b.c. using the provided data

Table 2. (continued)

	CONTRIBUTOR(S)	BASIS	SCHEME	DOMAIN DECOMPOSITION	MESH SIZE[1]	TORQUE[2]	REMARKS
{8}	Grimbert / Verry / El Ghazzani (2)	Euler	Clebsch, F.E. (Euler 3D)	Single block	1862 20-noded quadratic elem.	0.29090 (+13.73%)	Inlet b.c. using the provided data and adjusting the flowrate
{9}	Liess / Ecer	Euler	Clebsch, F.E. (PASSAGE)	"Block by block"	$59 \times 10 \times 10$	0.26300 (+2.83%)	
{10}	Lymberopoulos / Giannakoglou / Chaviaropoulos / Papailiou	Potential	F.D. / F.V.	Single block	$70 \times 17 \times 10$	0.20015 (-21.75%)	Adjusted data for potential assumption
{11}	Nagafuji / Suzuki / Kobayashi / Taniguchi (1)	Potential	F.E.	Single block	$36 \times 17 \times 7$	0.27220 (+6.42%)	Quasi-free vortex flow at the inlet
{12}	Nagafuji / Suzuki / Kobayashi / Taniguchi (2)	Potential	F.E.	Single block	$36 \times 17 \times 7$	0.29370 (+14.83%)	Free vortex flow at the inlet
{13}	Vu / Shyy	N.S. (Laminar, Re = 800)	Semi-impl. F.V.	Single block	$59 \times 19 \times 13$[4]	0.25452 (-0.49%)	

[1]All the contributors have used H-H mesh topology.

[2]The relative error with respect to the experimental torque $t^* = 0.25577$ is given between brackets.

[3]The effect of mesh size was investigated; the mesh indicated is the finest used.

[4]Determined from the figure of the mesh in the paper.

Table 3. Draft tube calculations - Brief overview of the contributions.

CONTRIBUTOR(S)	BASIS	SCHEME	MESH TOPOLOGY	MESH SIZE	RECIRCULATION	DOWNSTREAM B.C.	REMARKS
Kubota / Toshigami / Suzuki	N.S. $\nu_t = const$ (2 values)	F.E.	Unstructured, 10-nodes tetrahedron	9729 elem.	No	p = const, Neumann conditions on velocity components	$\nu_t = const_1$ in core regions, $\nu_t = const_2$ in wall regions
Lazzaro / Riva	N.S. $k\text{-}\varepsilon \Rightarrow \nu_t$	F.E. (FIDAP)	Unstructured, 8-nodes brick	6948 elem.	Yes	Default b.c. provided by FIDAP	$Re = 1.08 \times 10^6$ turbulent Wall functions used
Liess / Ecer	Euler	Expl. F.V., artificial compressibility	Structured, H-H	5082 points[1]	No	Characteristic non-reflecting b.c. forcing p to converge to p_{dtout}	Convergence problems for higher grid density
Ruprecht (1)	N.S. $k\text{-}\varepsilon \Rightarrow \nu_t$	F.E. (FENFLOSS)	Unstructured, 8-nodes brick	12384 elem.	Yes	$\frac{\partial u_i}{\partial x} = 0$ $\frac{\partial k}{\partial x} = \frac{\partial \varepsilon}{\partial x} = 0$ $p = p_{ref}$	Wall functions used
Ruprecht (2)	N.S. $k\text{-}\varepsilon \Rightarrow \nu_t$	F.E. (FENFLOSS)	Unstructured, 8-nodes brick	12384 elem.	Yes	$\frac{\partial u_i}{\partial x} = 0$ $\frac{\partial k}{\partial x} = \frac{\partial \varepsilon}{\partial x} = 0$ $p = p_{dtout}$	Wall functions used

Table 3. (continued)

CONTRIBUTOR(S)	BASIS	SCHEME	MESH TOPOLOGY	MESH SIZE	RECIRCULATION	DOWNSTREAM B.C.	REMARKS
Takagi / Tanabe / Ikegawa / Mukai / Sato	N.S. $k\text{-}\varepsilon \Rightarrow \nu_t$	F.E. (ABMAC)	Structured, Hexahedrons	45625 points[1]	Yes, + transverse vortices in detached zones	$\frac{\partial k}{\partial x} = \frac{\partial \varepsilon}{\partial x} = 0$ $n_i \{\pi - (\upsilon_t + \upsilon)(U_{i,j} + U_{j,i})\} = 0$	Wall functions used
Vu / Shyy	N.S. $k\text{-}\varepsilon \Rightarrow \nu_t$	Semi-impl. F.V.	Structured, H-H	63125 points[1] + 38367 points	Yes, + a pair of counterrotating longitudinal vortices in detached zone	staggered grid $\Rightarrow$ no b.c. for p; zero gradients streamwise for all other variables	Smoothed + original geometry Wall functions used

[1]The effect of mesh size was investigated; the mesh indicated is the finest used.

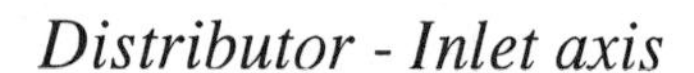

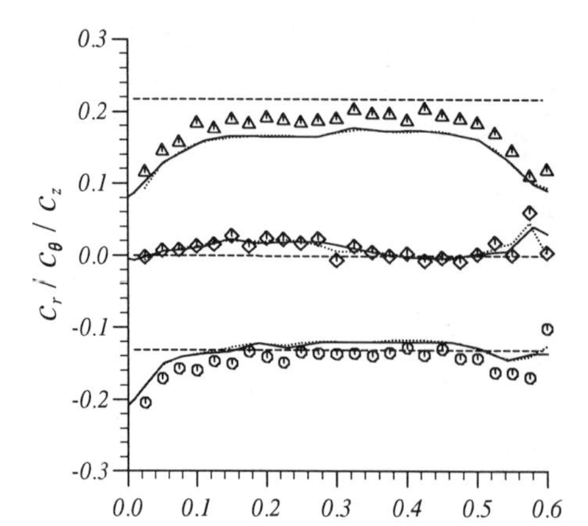

Experiment - c_r
Experiment - c_θ
Experiment - c_z

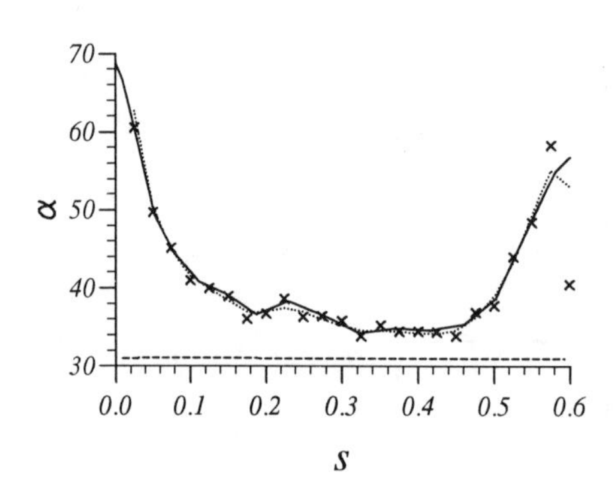

× Experiment - α

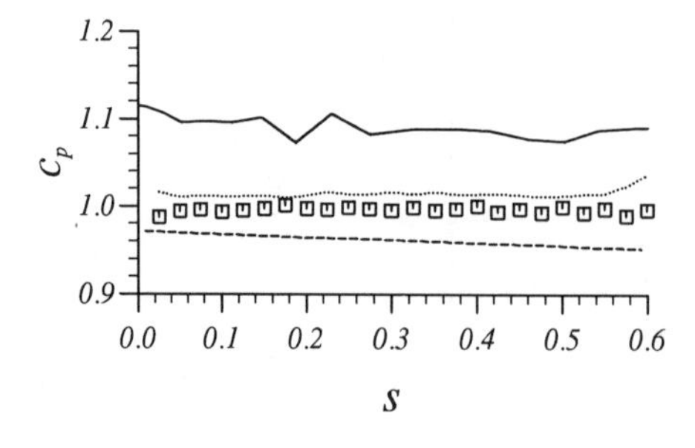

Experiment - c_p

Fig. 1. Velocity components, flow angle and pressure coefficient. Comparison between the experimental data and the imposed/computed conditions at the distributor inlet axis.

—— *Bottaro/Drotz/Gamba/Sottas/Neury*
········ *Eliasson*
----- *Goede*

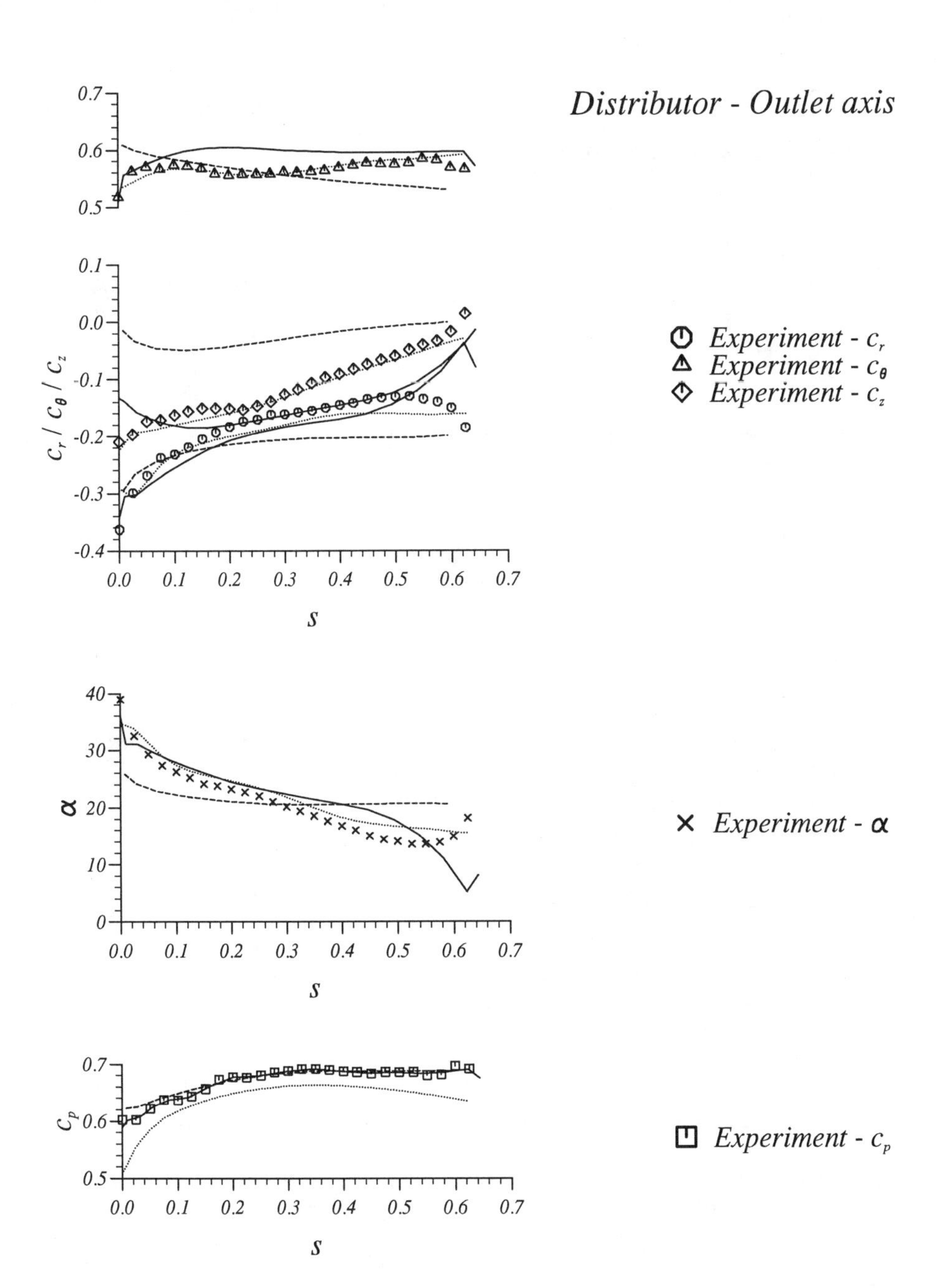

Fig. 2. Velocity components, flow angle and pressure coefficient. Comparison between the experimental data and the imposed/computed conditions at the distributor outlet axis.

—— *Bottaro/Drotz/Gamba/Sottas/Neury*
········ *Eliasson*
----- *Goede*

Distributor - Inlet axis

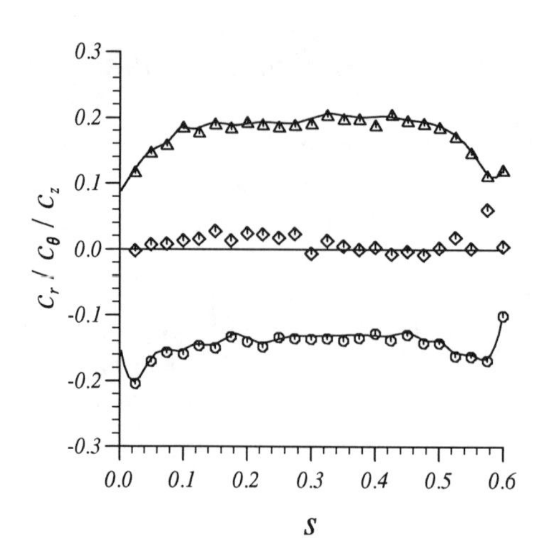

Experiment - c_r
Experiment - c_θ
Experiment - c_z

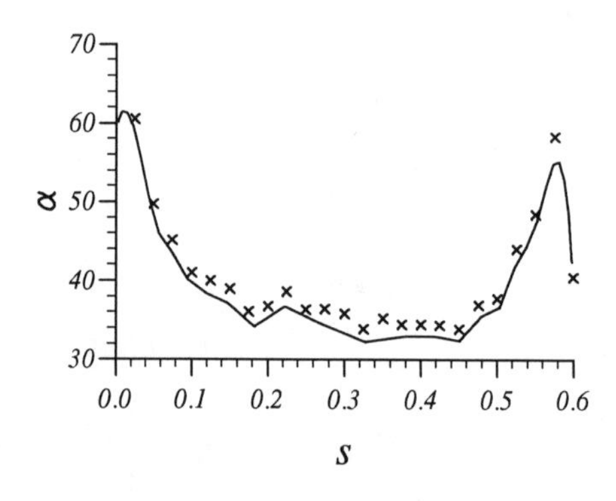

Experiment - α

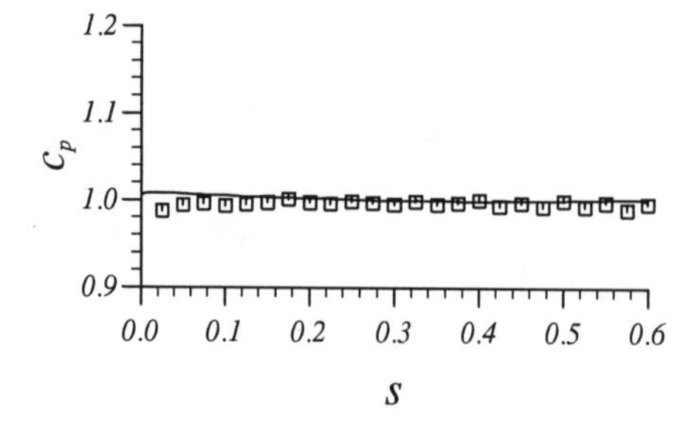

Experiment - c_p

Fig. 3. Velocity components, flow angle and pressure coefficient. Comparison between the experimental data and the imposed/computed conditions at the distributor inlet axis.

— *Nagafuji/Suzuki/Kobayashi/Taniguchi*

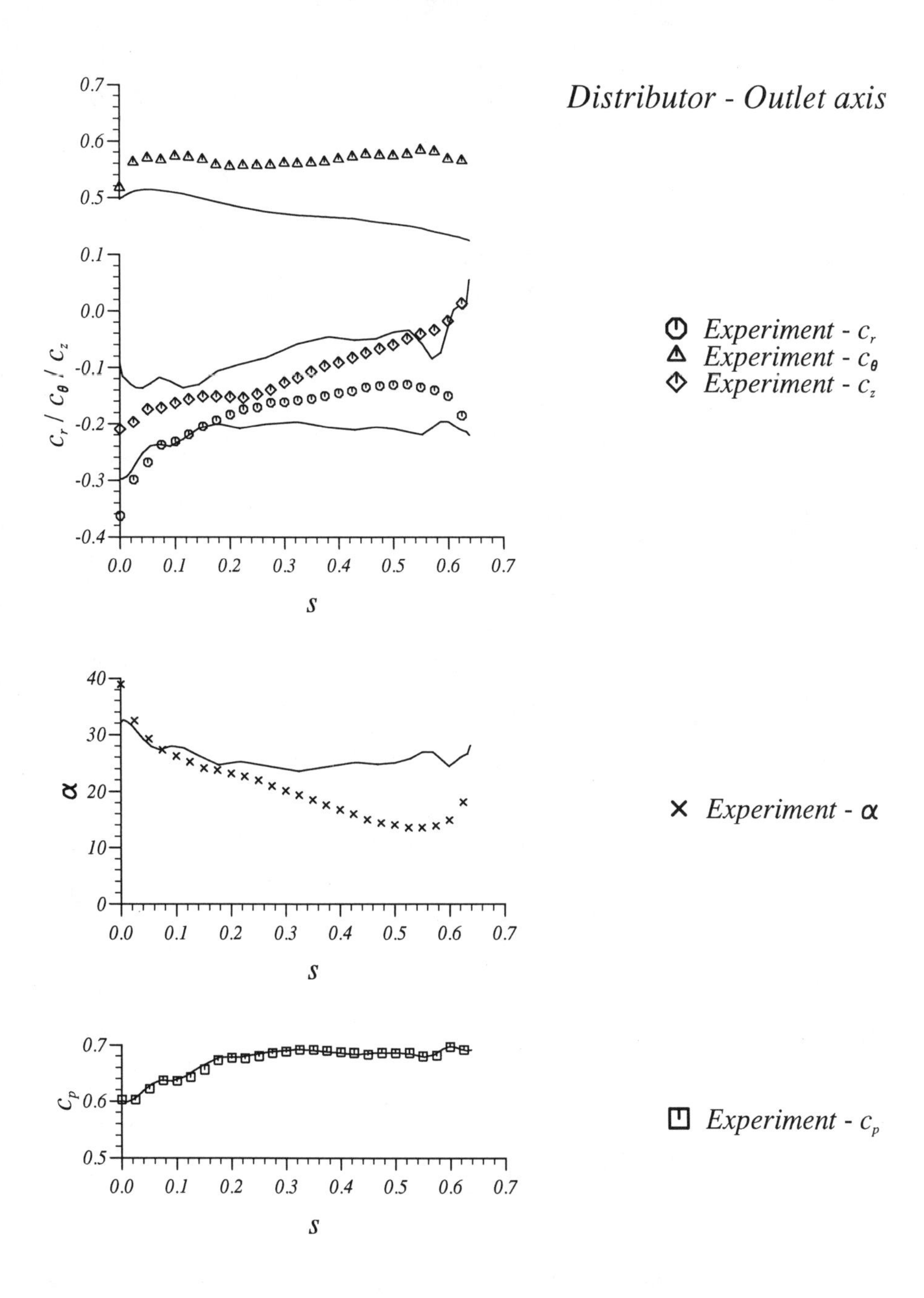

Fig. 4. Velocity components, flow angle and pressure coefficient. Comparison between the experimental data and the imposed/computed conditions at the distributor outlet axis.

—— Nagafuji/Suzuki/Kobayashi/Taniguchi

Stay vane - Z = 0

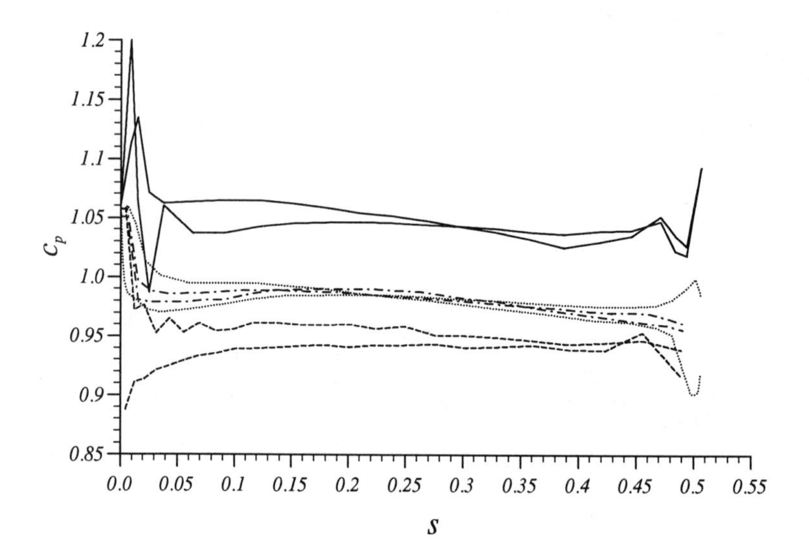

Fig. 5. Pressure distribution at midheight of the distributor blades. Comparison between the computed results from

—— Bottaro/Drotz/Gamba/Sottas/Neury
········ Eliasson
----- Goede
-·- Nagafuji/Suzuki/Kobayashi/Taniguchi

Wicket gate - Z = 0

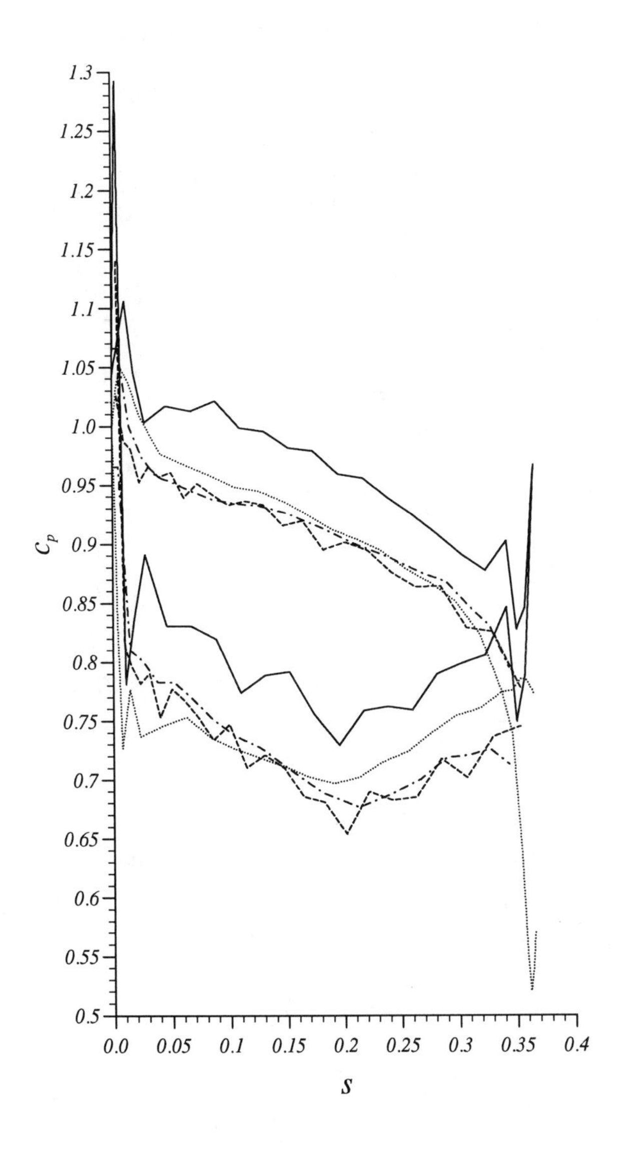

Fig. 5. (Contd.)

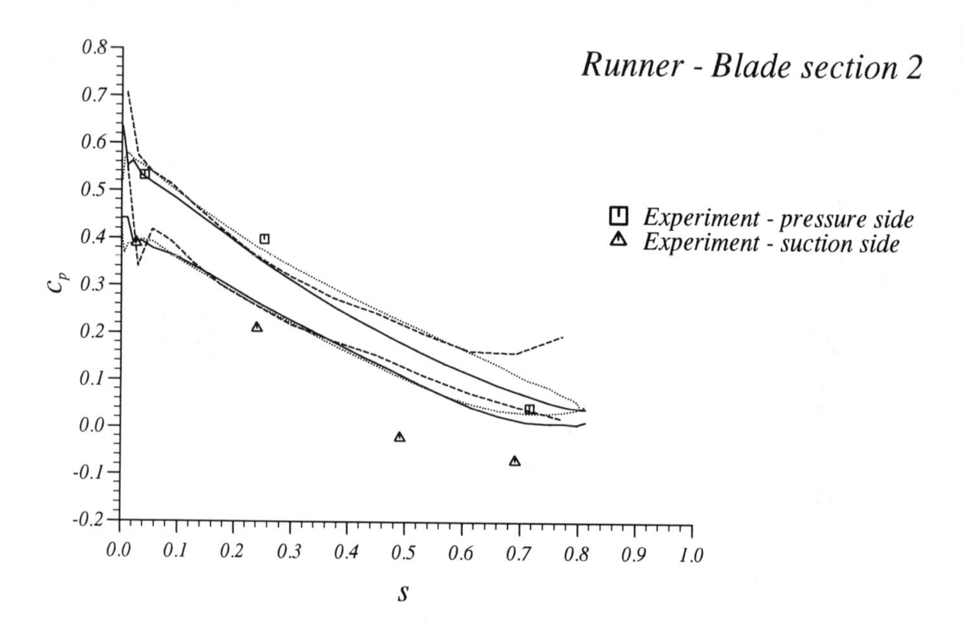

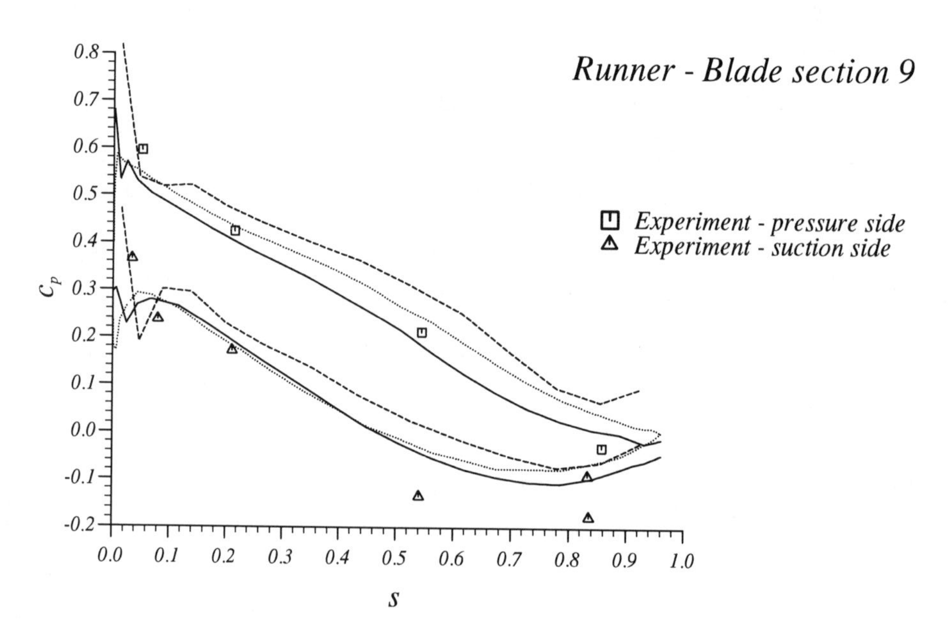

Fig. 6. Pressure distribution on three sections of the runner blade. Comparison between the experimental data and the computed results from

—— *Billdal/Jacobsen/Bratsberg/Andersson/Brekke {2}*
.... *Eliasson {5}*
----- *Goede {6}*

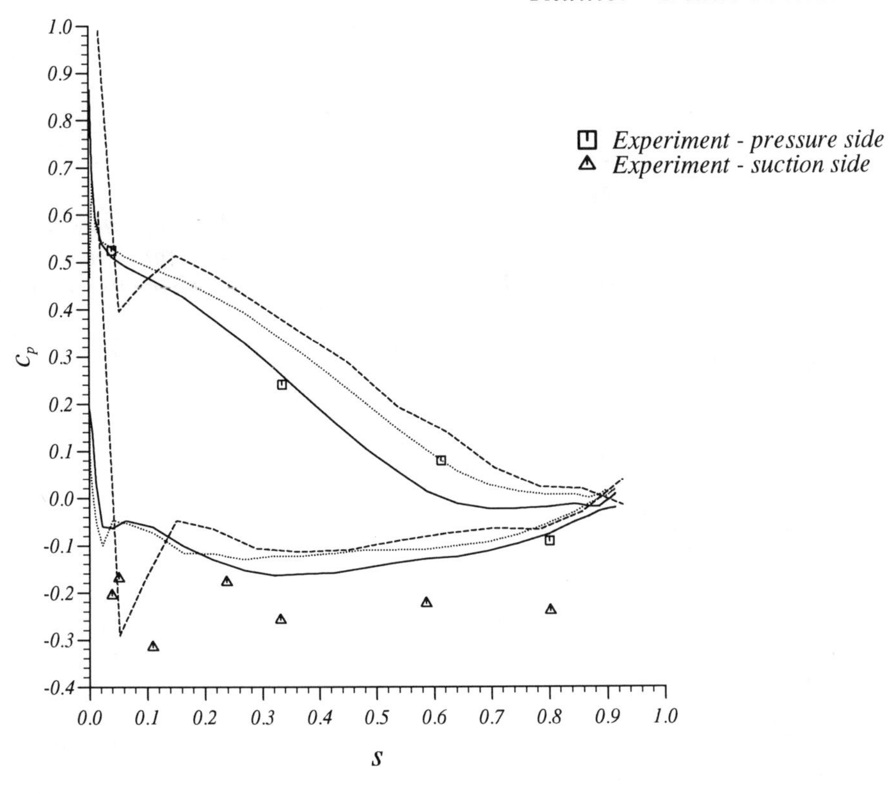
Runner - Blade section 15
Experiment - pressure side
Experiment - suction side
C_p
1.0
0.9
0.8
0.7
0.6
0.5
0.4
0.3
0.2
0.1
0.0
-0.1
-0.2
-0.3
-0.4
0.0
0.1
0.2
0.3
0.4
0.5
0.6
0.7
0.8
0.9
1.0
s

Fig. 6. (Contd.)

Runner - Inlet axis

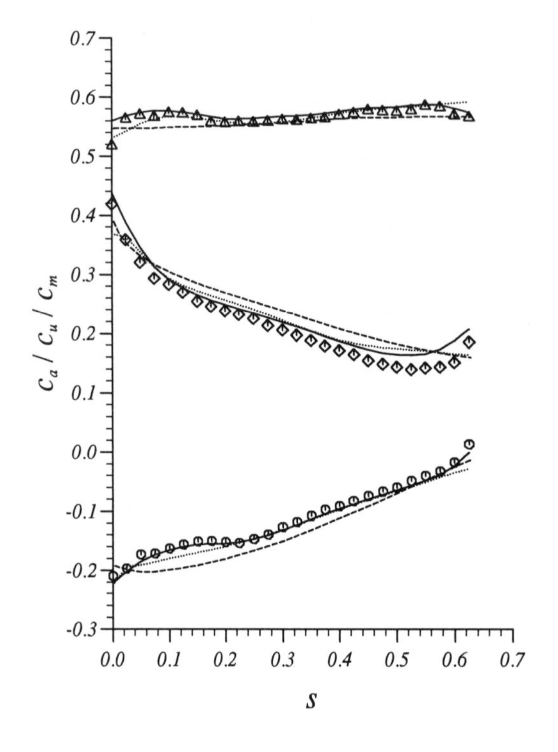

Experiment - c_a
Experiment - c_u
Experiment - c_m

Fig. 7. Velocity components, flow angles and pressure coefficient. Comparison between the experimental data and the imposed/computed conditions at the runner inlet axis.

—— *Billdal/Jacobsen/Bratsberg/Andersson/Brekke {2}*
········ *Eliasson {5}*
----- *Goede {6}*

Runner - Inlet axis

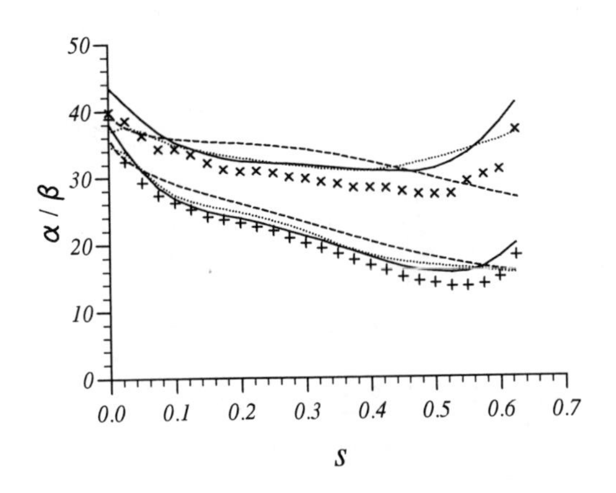

+ *Experiment* - α
× *Experiment* - β

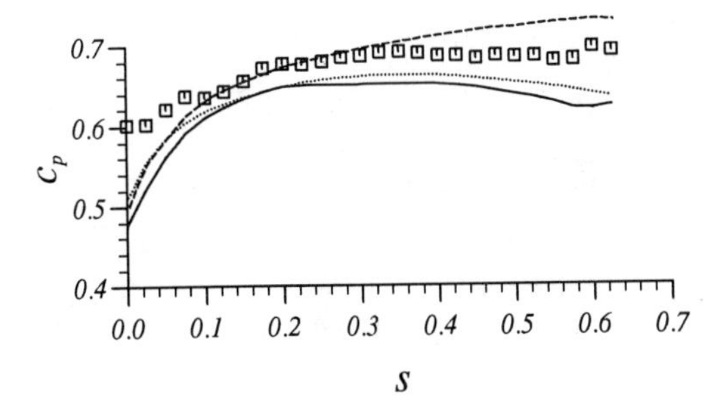

□ *Experiment* - c_p

Fig. 7. (Contd.)

Runner - Middle axis

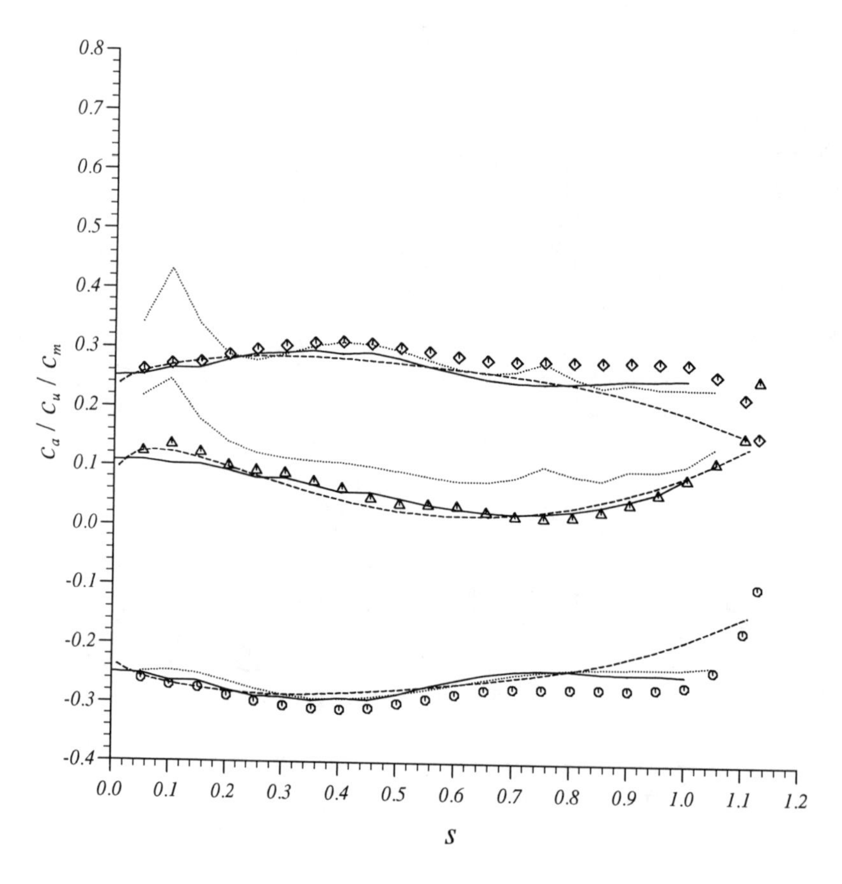

Experiment - c_a
Experiment - c_u
Experiment - c_m

Fig. 8. Velocity components, flow angles and pressure coefficient on the runner middle axis. Comparison between the experimental data and the computed results from

— Billdal/Jacobsen/Bratsberg/Andersson/Brekke {2}
······ Eliasson {5}
----- Goede {6}

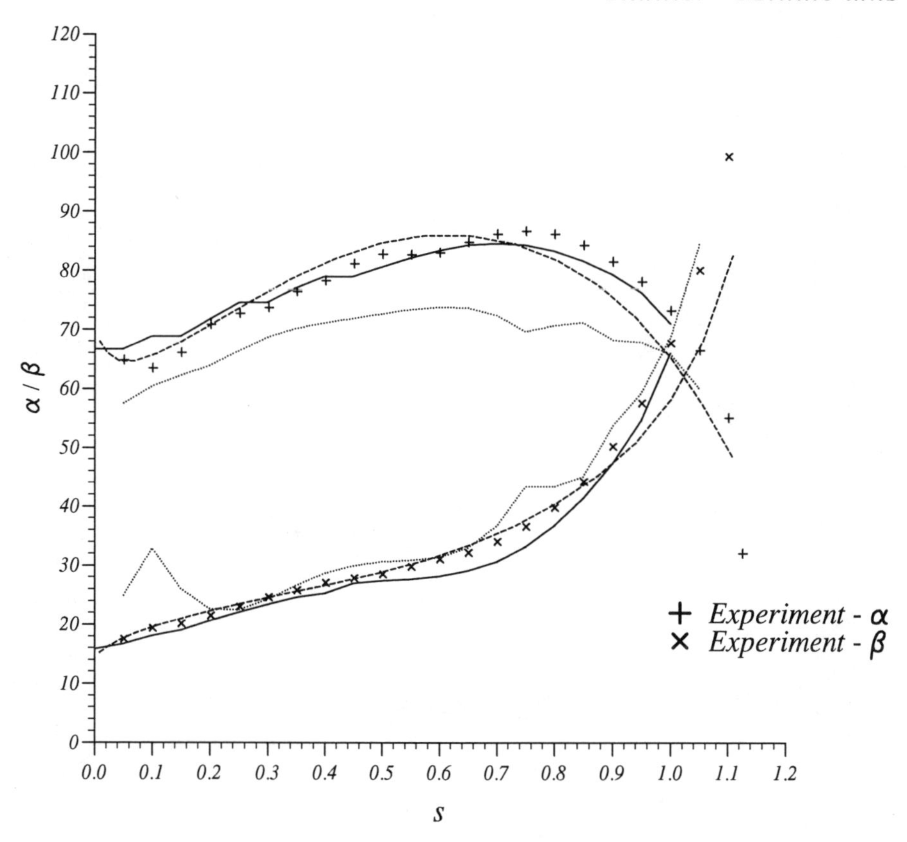
Runner - Middle axis
120
110
100
90
80
70
60
50
40
30
20
10
0
α / β
0.0
0.1
0.2
0.3
0.4
0.5
0.6
0.7
0.8
0.9
1.0
1.1
1.2
s
+ Experiment - α
× Experiment - β

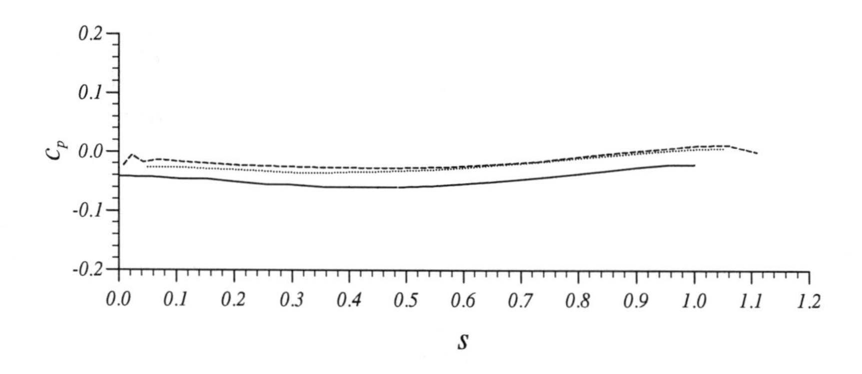
0.2
0.1
0.0
-0.1
-0.2
c_p
0.0
0.1
0.2
0.3
0.4
0.5
0.6
0.7
0.8
0.9
1.0
1.1
1.2
s

Fig. 8. (Contd.)

Runner - Outlet axis

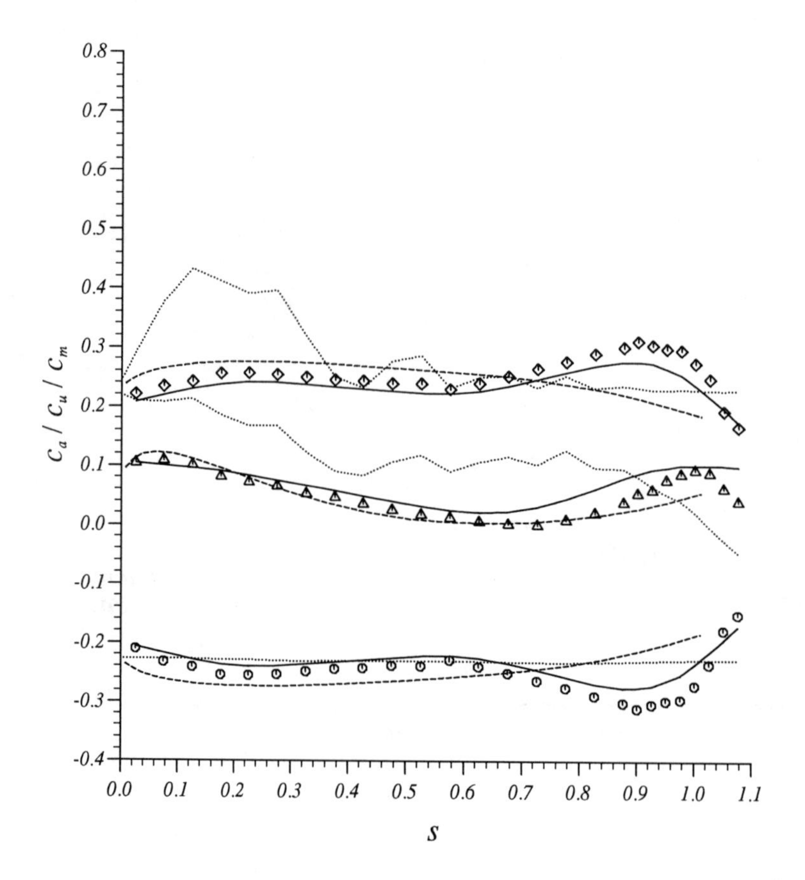

Fig. 9. *Velocity components, flow angles and pressure coefficient on the runner outlet axis. Comparison between the experimental data and the computed results from*
—— *Billdal/Jacobsen/Bratsberg/Andersson/Brekke {2}*
········ *Eliasson {5}*
----- *Goede {6}*

Runner - Outlet axis

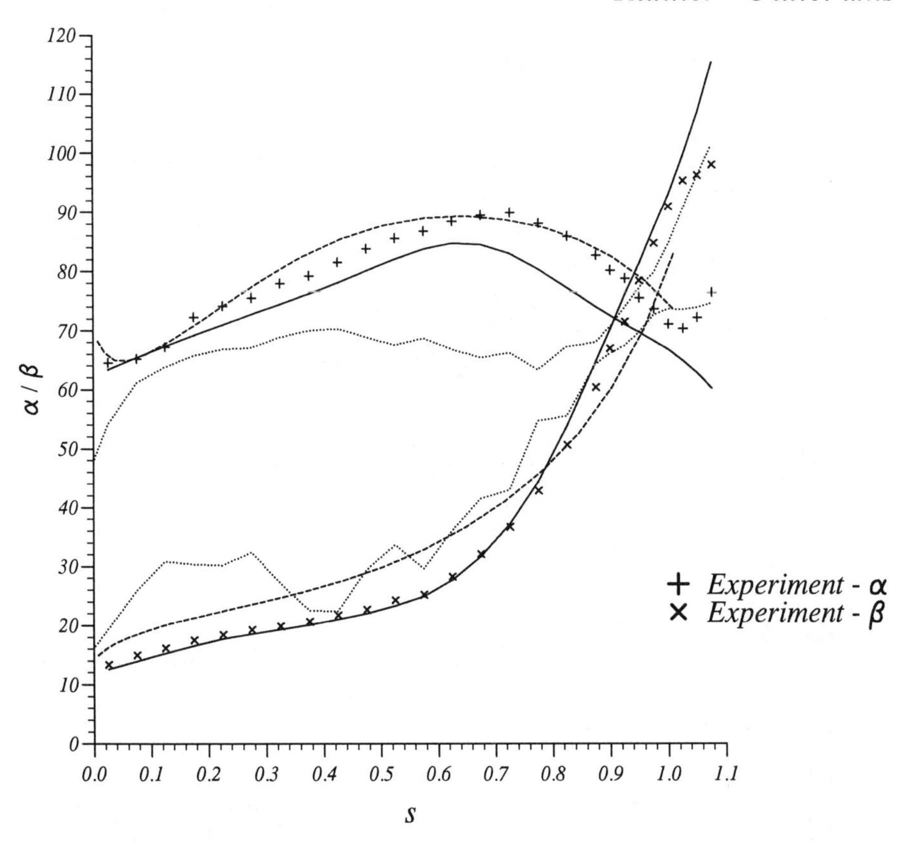

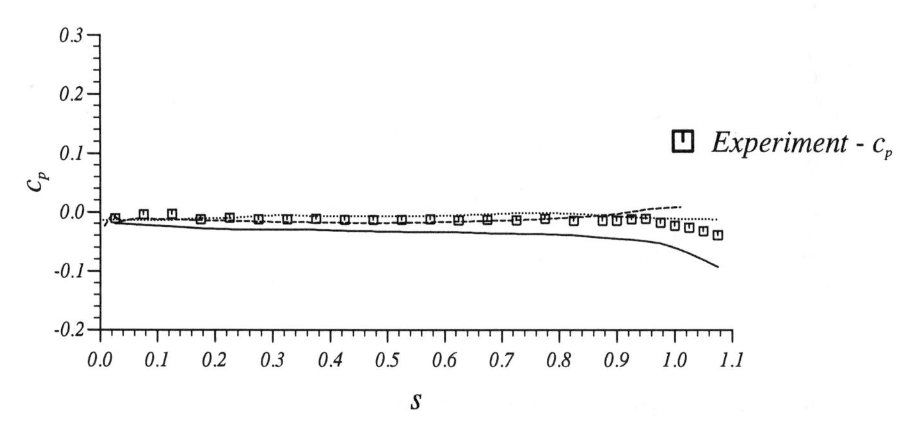

Fig. 9. (Contd.)

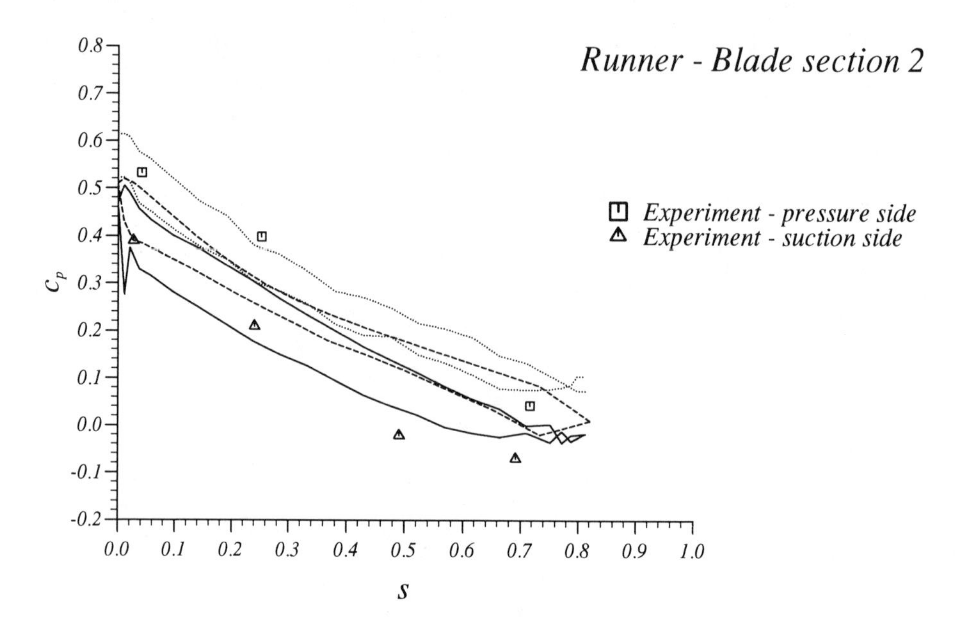

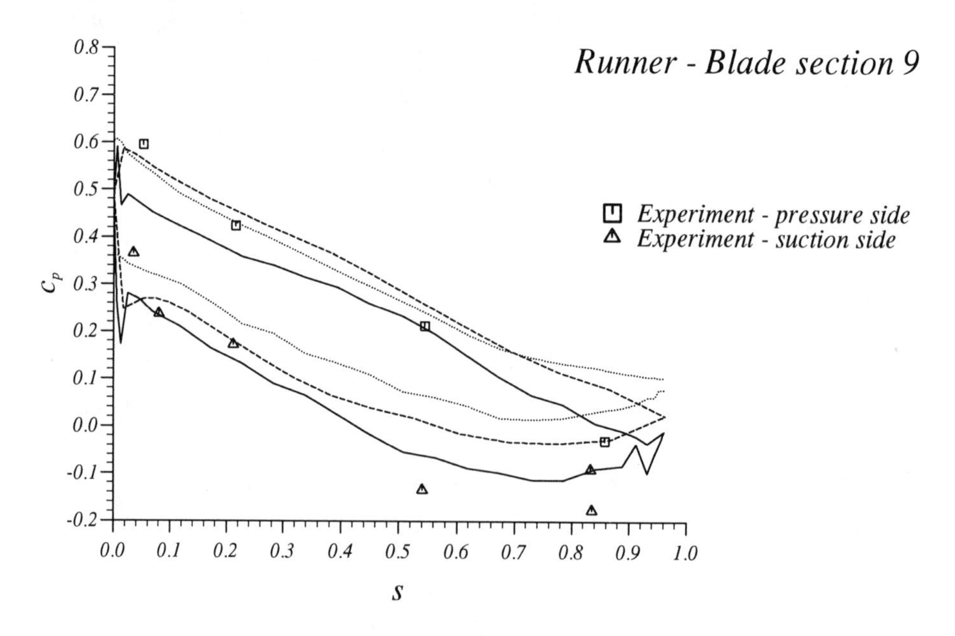

Fig. 10. Pressure distribution on three sections of the runner blade. Comparison between the experimental data and the computed results from

—— Bottaro/Drotz/Gamba/Sottas/Neury (1) {3}
········ Grimbert/Verry/El Ghazzani (1) {7}
----- Nagafuji/Suzuki/Kobayashi/Taniguchi (1) {11}

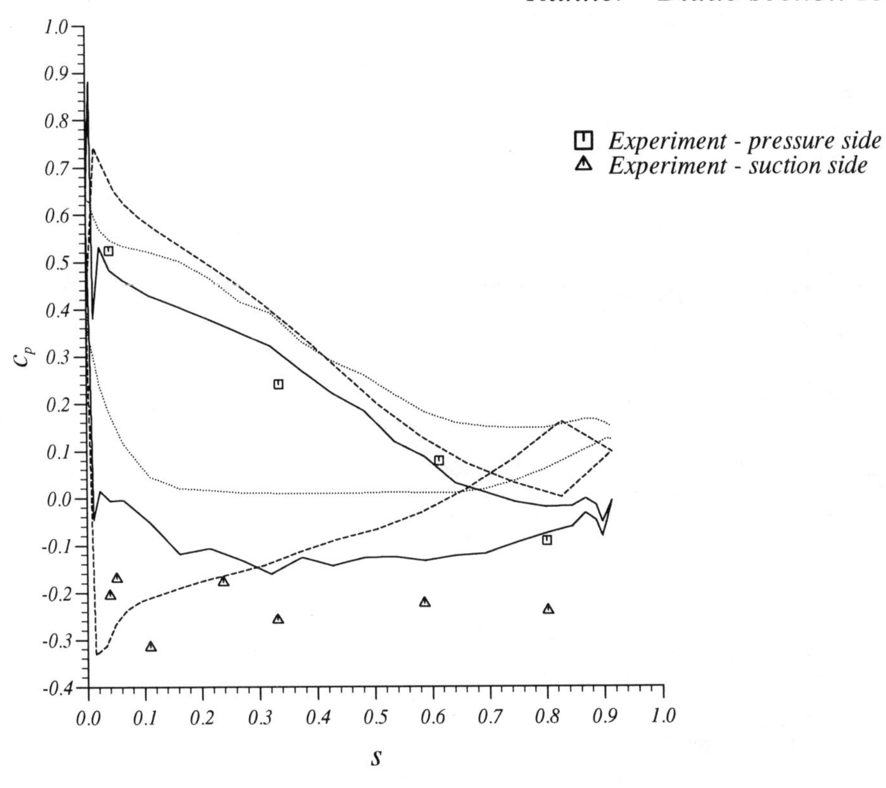
Runner - Blade section 15
Experiment - pressure side
Experiment - suction side
1.0
0.9
0.8
0.7
0.6
0.5
0.4
0.3
0.2
0.1
0.0
-0.1
-0.2
-0.3
-0.4
c_p
0.0
0.1
0.2
0.3
0.4
0.5
0.6
0.7
0.8
0.9
1.0
s

Fig. 10. (Contd.)

Runner - Inlet axis

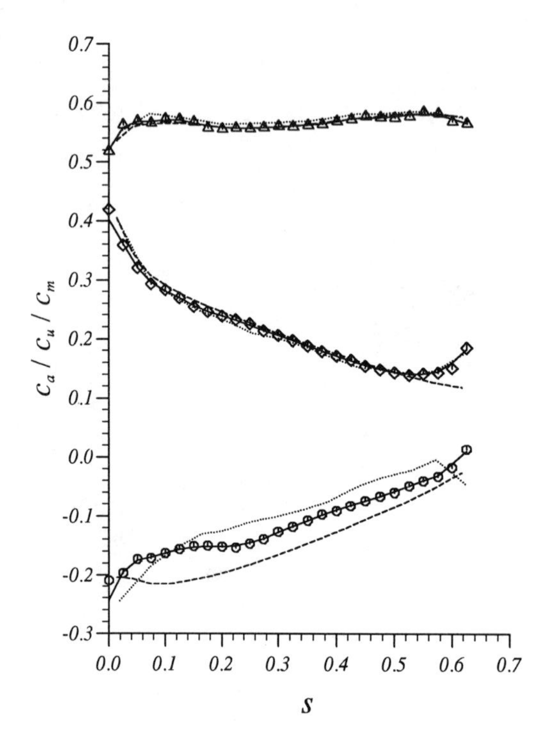

Experiment - c_a
Experiment - c_u
Experiment - c_m

Fig. 11. Velocity components, flow angles and pressure coefficient. Comparison between the experimental data and the imposed/computed conditions at the runner inlet axis.

—— *Bottaro/Drotz/Gamba/Sottas/Neury (1) {3}*
········ *Grimbert/Verry/El Ghazzani (1) {7}*
----- *Nagafuji/Suzuki/Kobayashi/Taniguchi (1) {11}*

Runner - Inlet axis

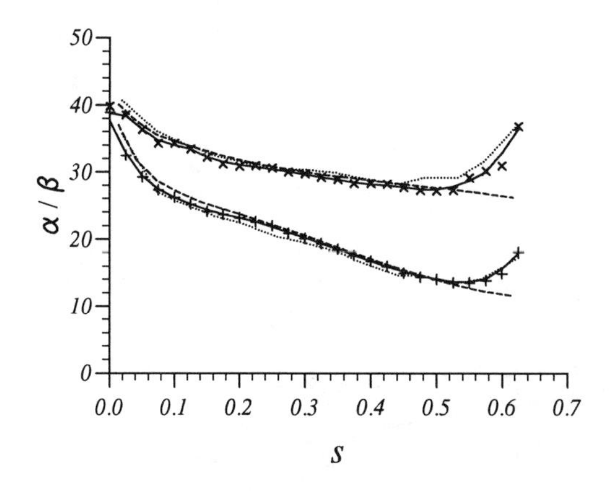

+ *Experiment* - α
× *Experiment* - β

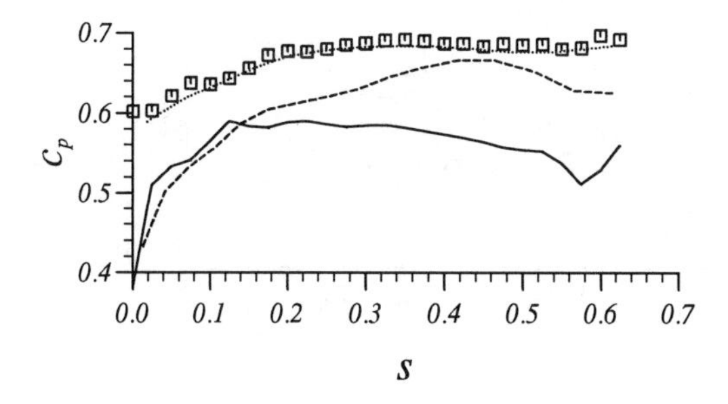

□ *Experiment* - c_p

Fig. 11. (Contd.)

Runner - Middle axis

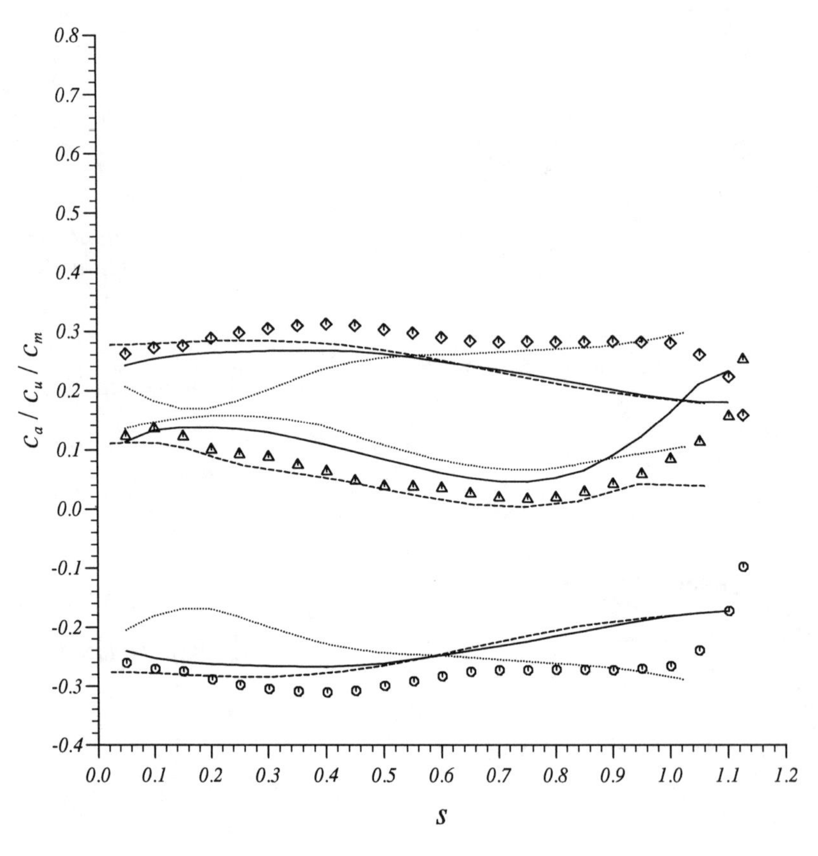

Fig. 12. Velocity components, flow angles and pressure coefficient on the runner middle axis. Comparison between the experimental data and the computed results from

—— *Bottaro/Drotz/Gamba/Sottas/Neury (1) {3}*
········ *Grimbert/Verry/El Ghazzani (1) {7}*
---- *Nagafuji/Suzuki/Kobayashi/Taniguchi (1) {11}*

Runner - Middle axis

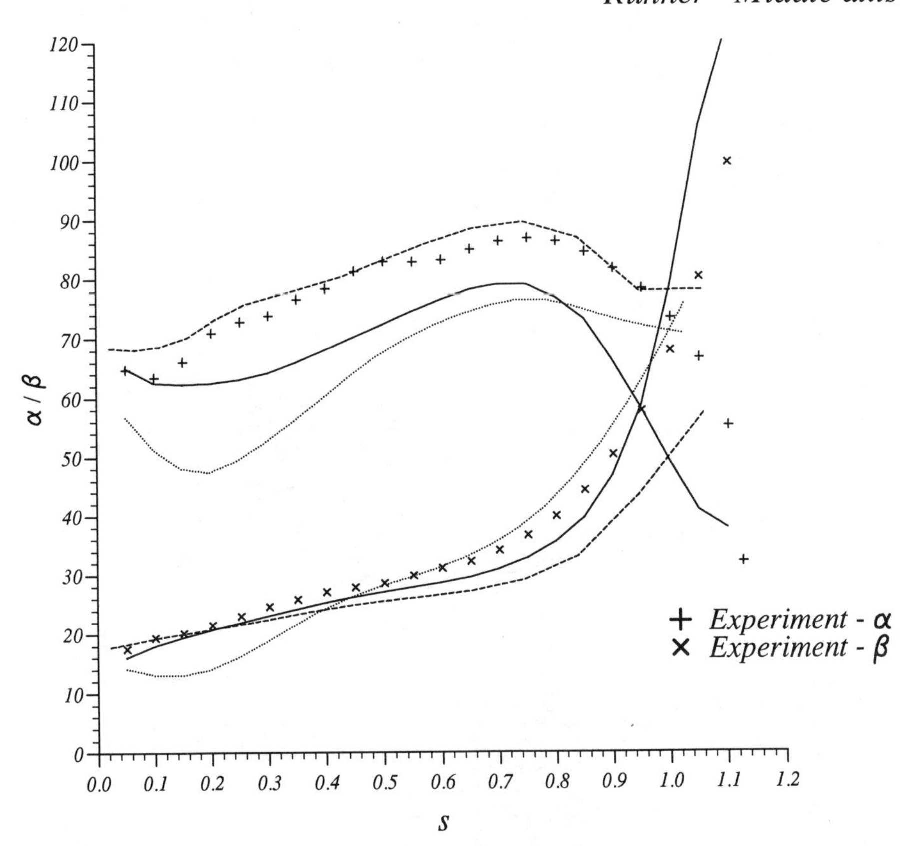

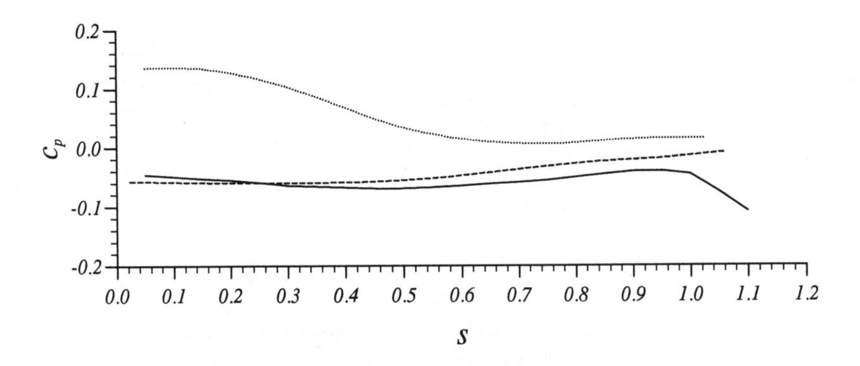

Fig. 12. (Contd.)

Runner - Outlet axis

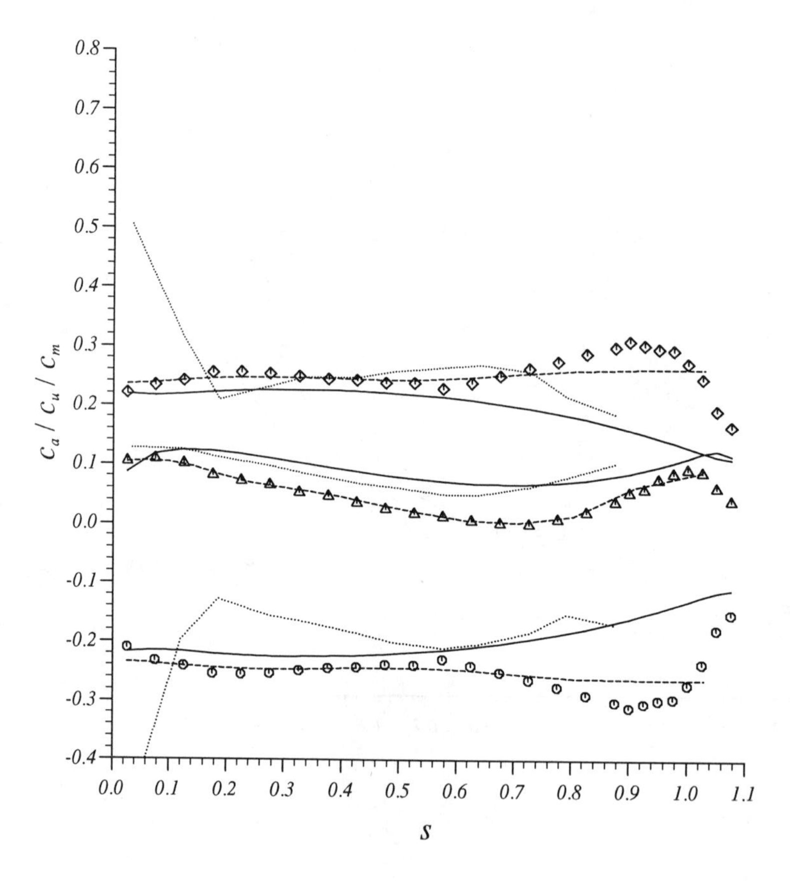

Experiment - c_a
Experiment - c_u
Experiment - c_m

Fig. 13. Velocity components, flow angles and pressure coefficient. Comparison between the experimental data and the imposed/computed conditions at the runner outlet axis.

—— *Bottaro/Drotz/Gamba/Sottas/Neury (1) {3}*
········ *Grimbert/Verry/El Ghazzani (1) {7}*
----- *Nagafuji/Suzuki/Kobayashi/Taniguchi (1) {11}*

Runner - Outlet axis

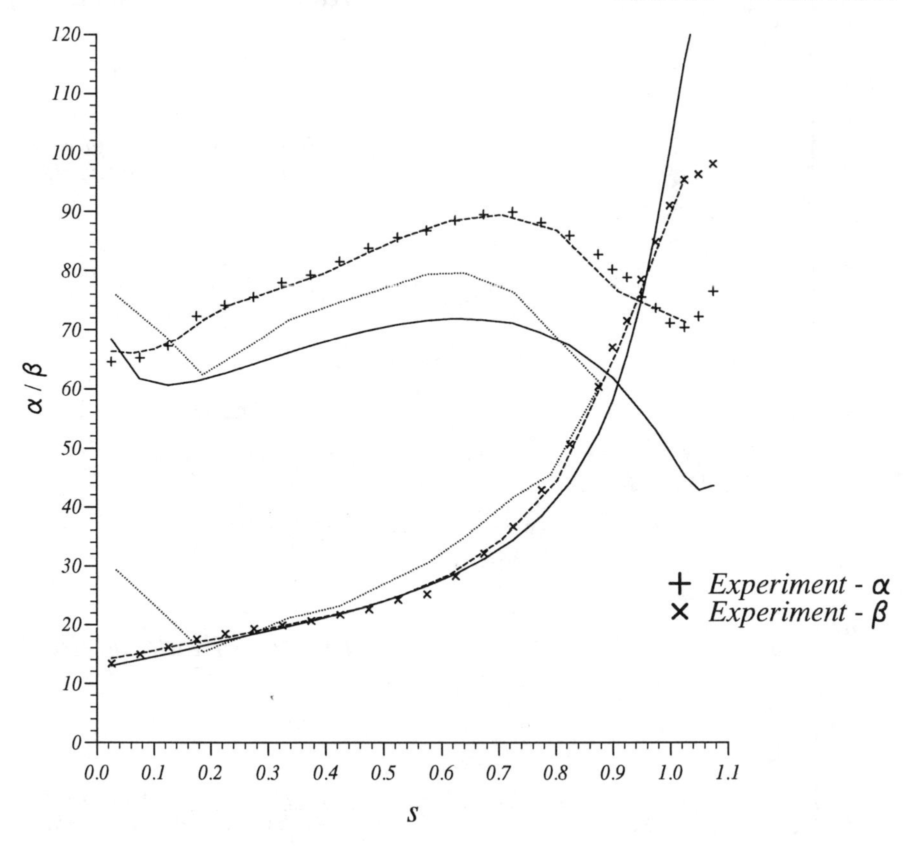

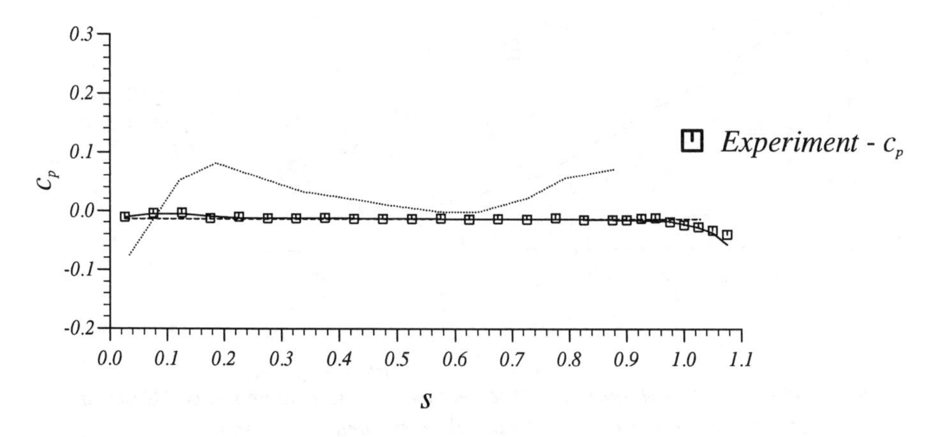

Fig. 13. (Contd.)

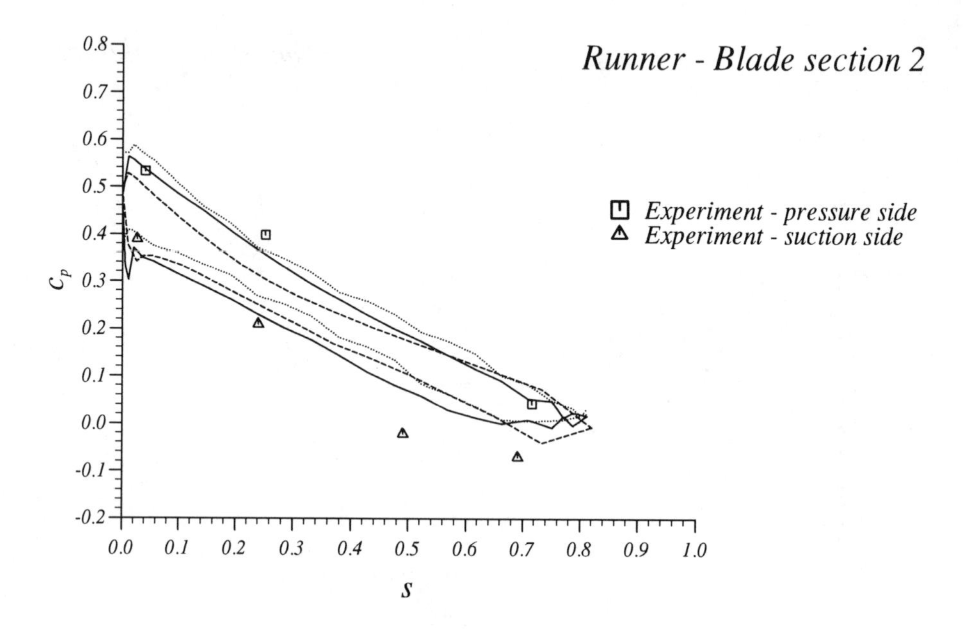

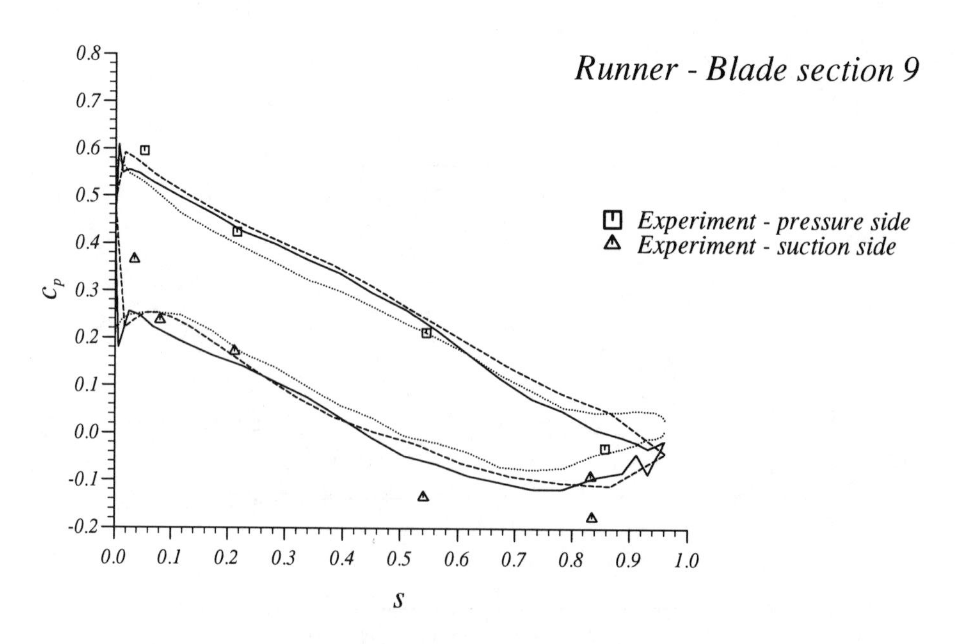

Fig. 14. Pressure distribution on three sections of the runner blade. Comparison between the experimental data and the computed results from

—— Bottaro/Drotz/Gamba/Sottas/Neury (2) {4}
.... Grimbert/Verry/El Ghazzani (2) {8}
---- Nagafuji/Suzuki/Kobayashi/Taniguchi (2) {12}

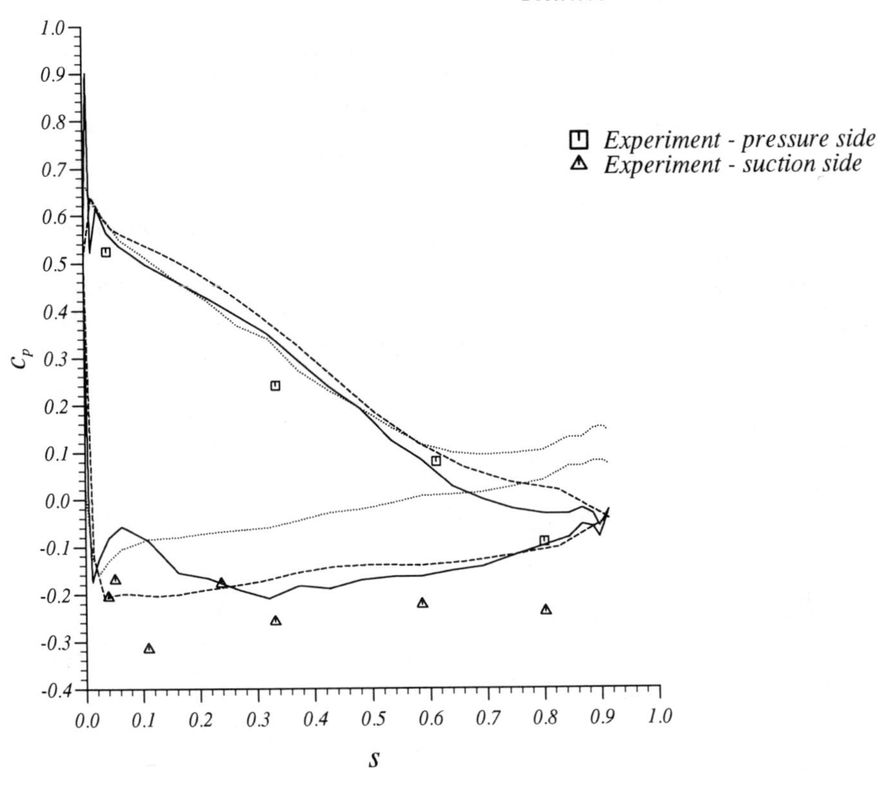
Runner - Blade section 15
Experiment - pressure side
Experiment - suction side
C_p
1.0
0.9
0.8
0.7
0.6
0.5
0.4
0.3
0.2
0.1
0.0
-0.1
-0.2
-0.3
-0.4
0.0
0.1
0.2
0.3
0.4
0.5
0.6
0.7
0.8
0.9
1.0
S

Fig. 14. (Contd.)

Runner - Inlet axis

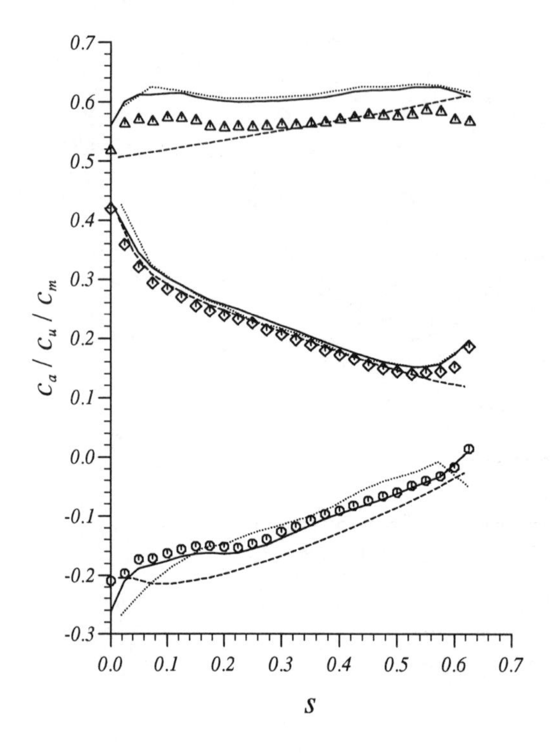

Fig. 15. Velocity components, flow angles and pressure coefficient. Comparison between the experimental data and the imposed/computed conditions at the runner inlet axis.

—— *Bottaro/Drotz/Gamba/Sottas/Neury (2) {4}*
········ *Grimbert/Verry/El Ghazzani (2) {8}*
----- *Nagafuji/Suzuki/Kobayashi/Taniguchi (2) {12}*

Runner - Inlet axis

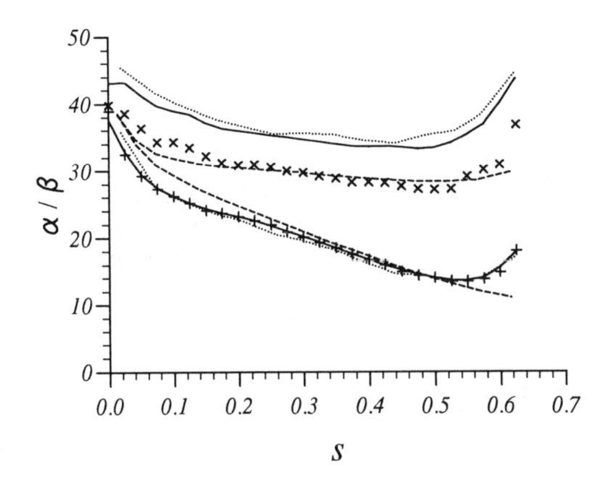

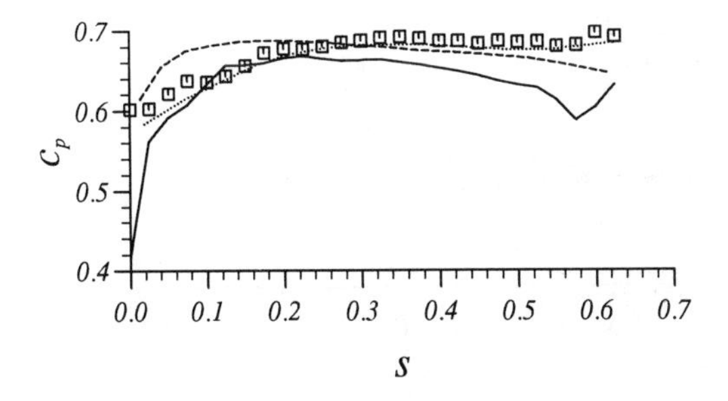

Fig. 15. (Contd.)

Runner - Middle axis

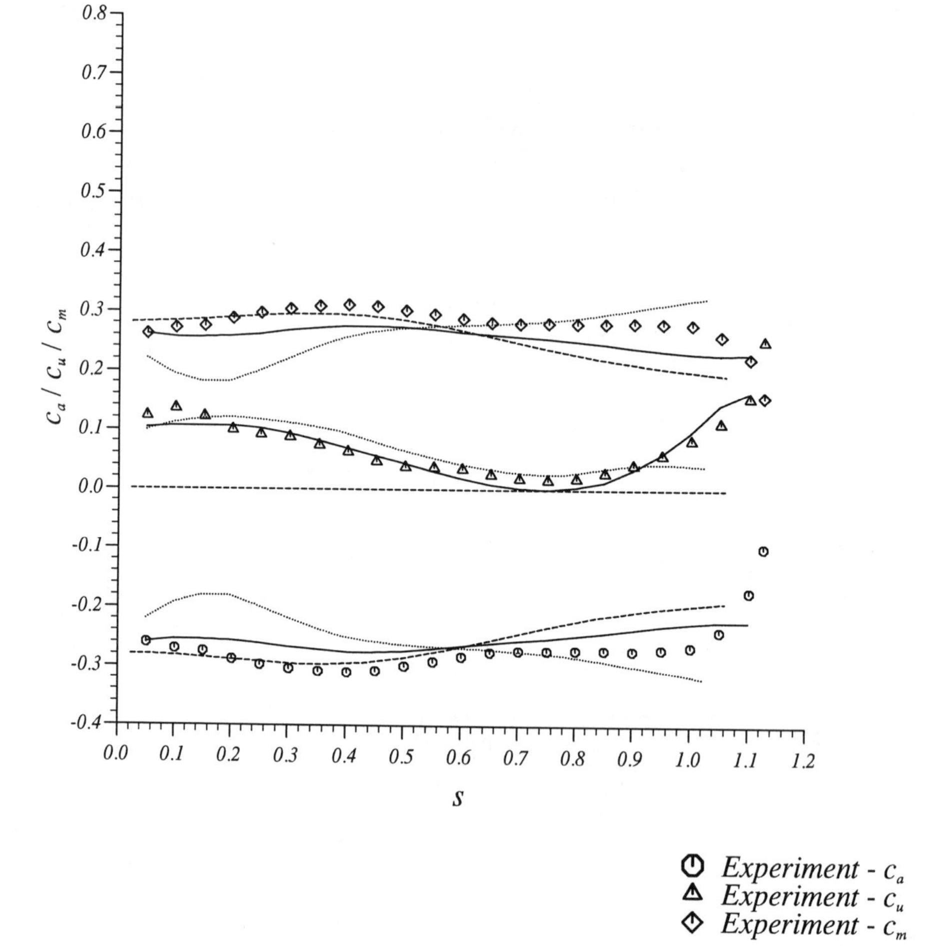

Fig. 16. Velocity components, flow angles and pressure coefficient on the runner middle axis. Comparison between the experimental data and the computed results from

—— Bottaro/Drotz/Gamba/Sottas/Neury (2) {4}
.... Grimbert/Verry/El Ghazzani (2) {8}
- - - - Nagafuji/Suzuki/Kobayashi/Taniguchi (2) {12}

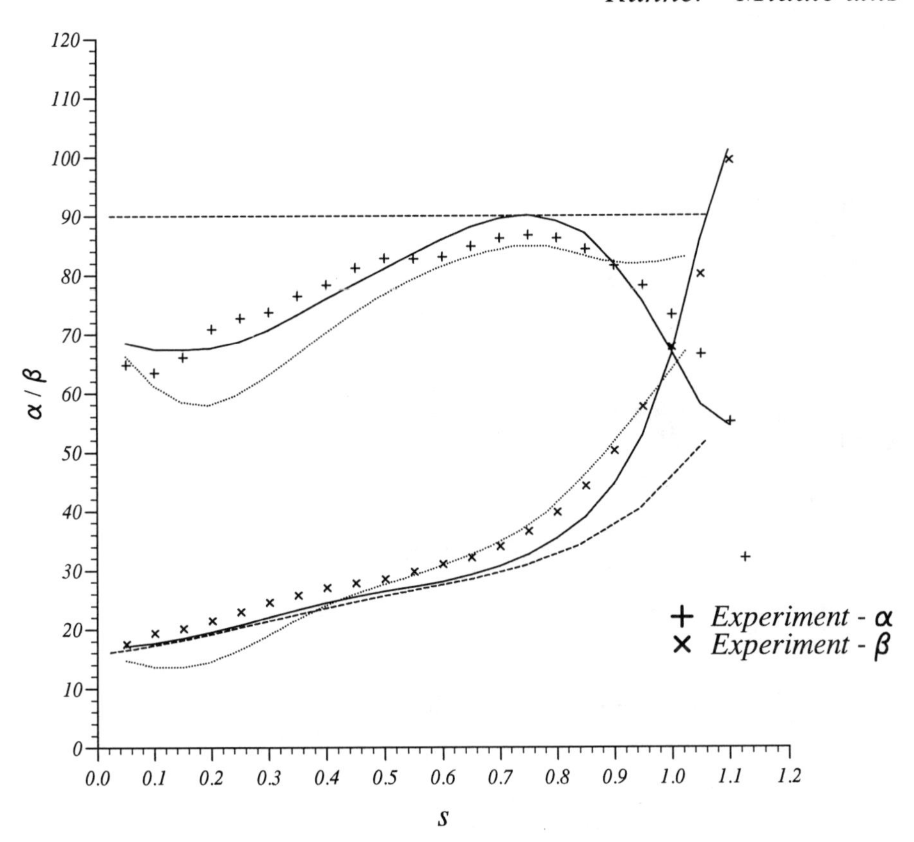
α / β
120
110
100
90
80
70
60
50
40
30
20
10
0
0.0 0.1 0.2 0.3 0.4 0.5 0.6 0.7 0.8 0.9 1.0 1.1 1.2
s
+ Experiment - α
× Experiment - β

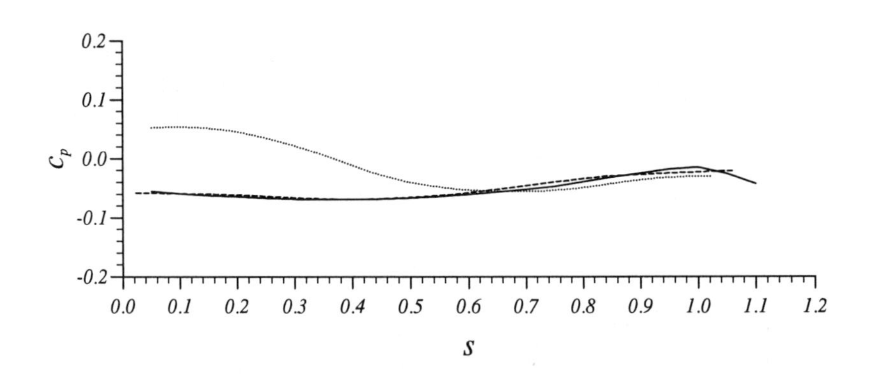
C_p
0.2
0.1
0.0
-0.1
-0.2
0.0 0.1 0.2 0.3 0.4 0.5 0.6 0.7 0.8 0.9 1.0 1.1 1.2
s

Fig. 16. (Contd.)

Runner - Outlet axis

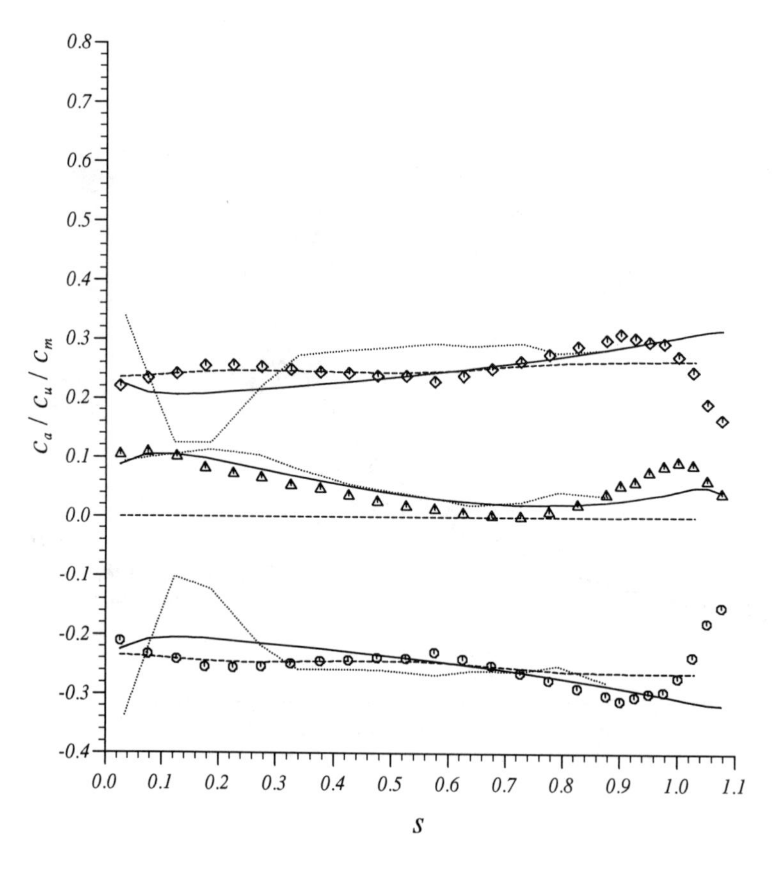

Fig. 17. Velocity components, flow angles and pressure coefficient. Comparison between the experimental data and the imposed/computed conditions at the runner outlet axis

—— *Bottaro/Drotz/Gamba/Sottas/Neury (2) {4}*
.... *Grimbert/Verry/El Ghazzani (2) {8}*
----- *Nagafuji/Suzuki/Kobayashi/Taniguchi (2) {12}*

Runner - Outlet axis

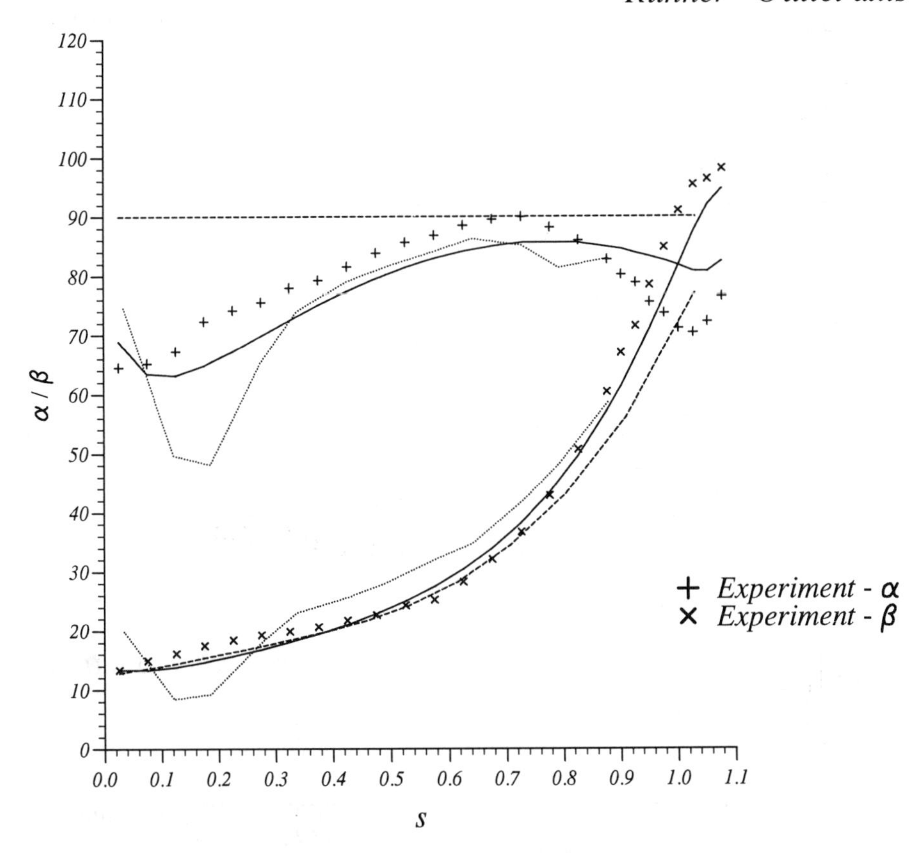

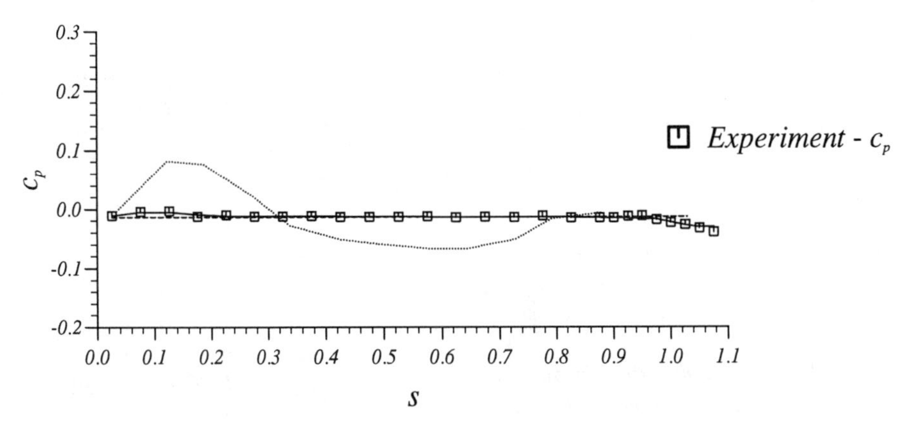

Fig. 17. (Contd.)

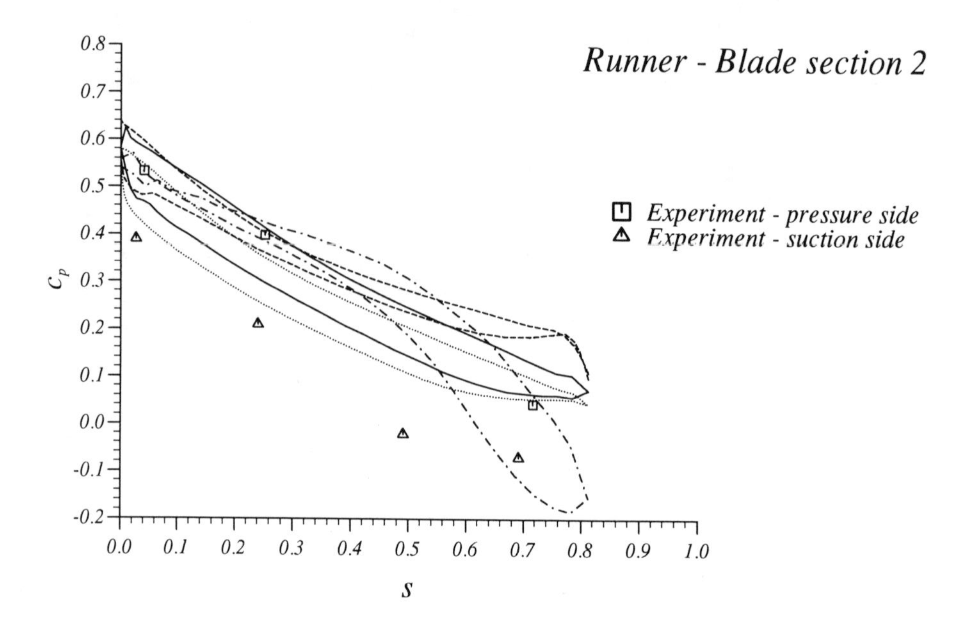

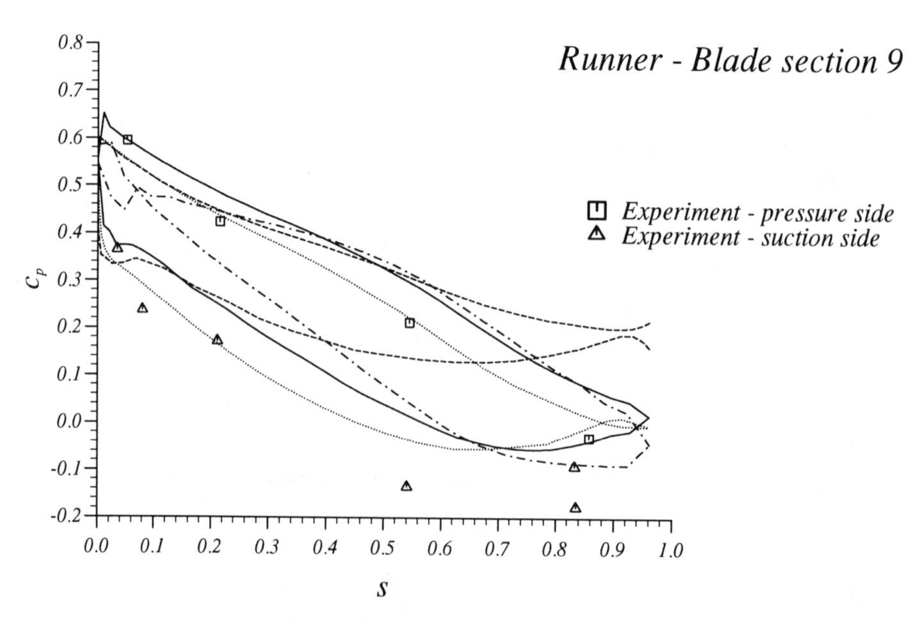

Fig. 18. Pressure distribution on three sections of the runner blade. Comparison between the experimental data and the computed results from

—— *Arakawa/Samejima/Kubota/Suzuki {1}*
········ *Liess/Ecer {9}*
----- *Lymberopoulos/Giannakoglou/Chaviaropoulos/Papailiou {10}*
-·- *Vu/Shyy {13}*

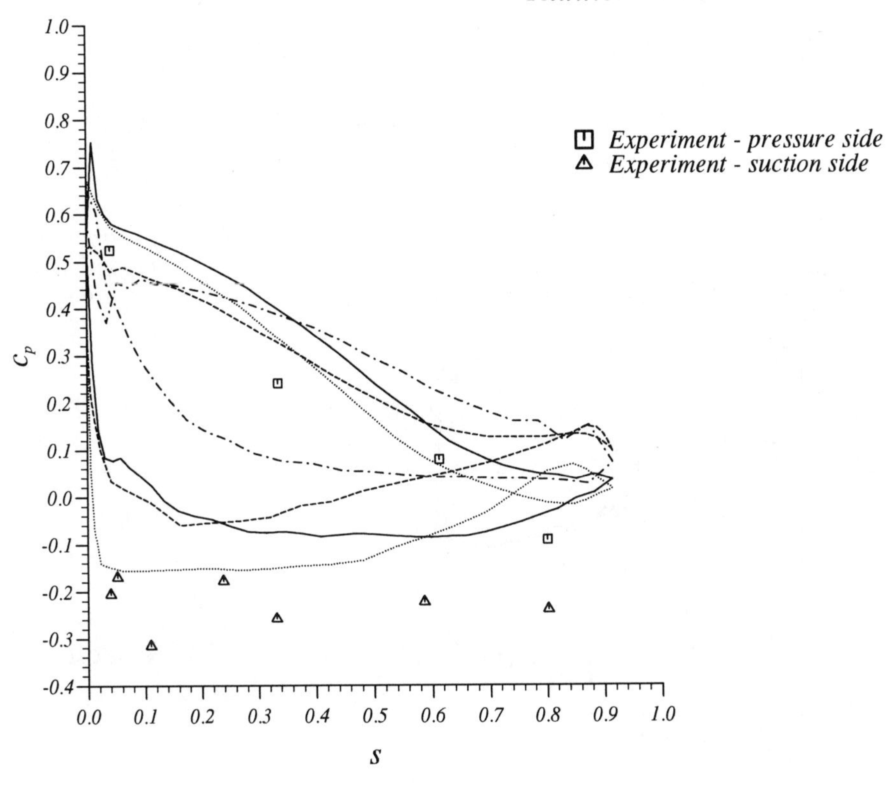
Runner - Blade section 15
Experiment - pressure side
Experiment - suction side
C_p
1.0
0.9
0.8
0.7
0.6
0.5
0.4
0.3
0.2
0.1
0.0
-0.1
-0.2
-0.3
-0.4
0.0
0.1
0.2
0.3
0.4
0.5
0.6
0.7
0.8
0.9
1.0
S

Fig. 18. (Contd.)

Runner - Inlet axis

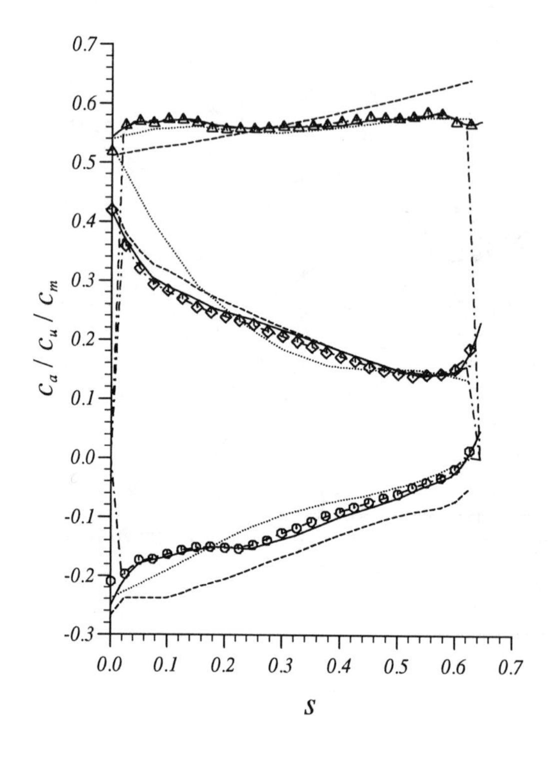

Experiment - c_a
Experiment - c_u
Experiment - c_m

Fig. 19. Velocity components, flow angles and pressure coefficient. Comparison between the experimental data and the imposed/computed conditions at the runner inlet axis.

—— *Arakawa/Samejima/Kubota/Suzuki {1}*
······· *Liess/Ecer {9}*
----- *Lymberopoulos/Giannakoglou/Chaviaropoulos/Papailiou {10}*
-·- *Vu/Shyy {13}*

Runner - Inlet axis

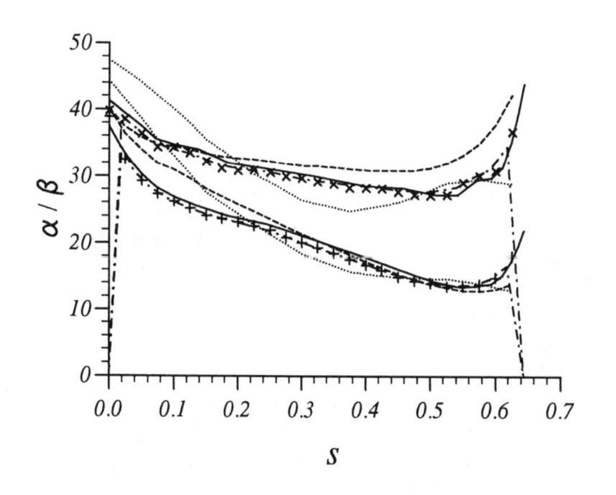

+ *Experiment* - α
× *Experiment* - β

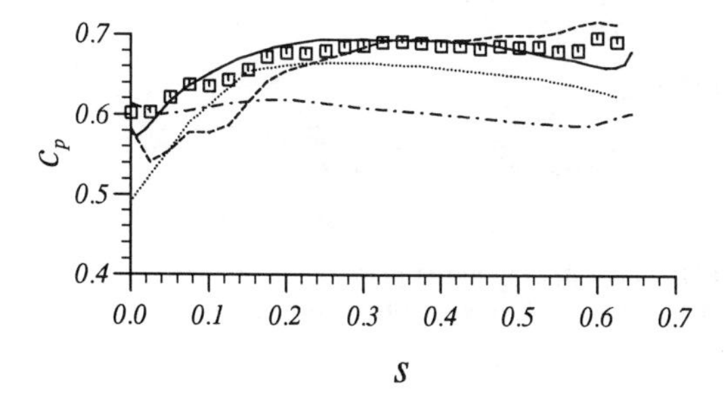

⊡ *Experiment* - c_p

Fig. 19. (Contd.)

Runner - Middle axis

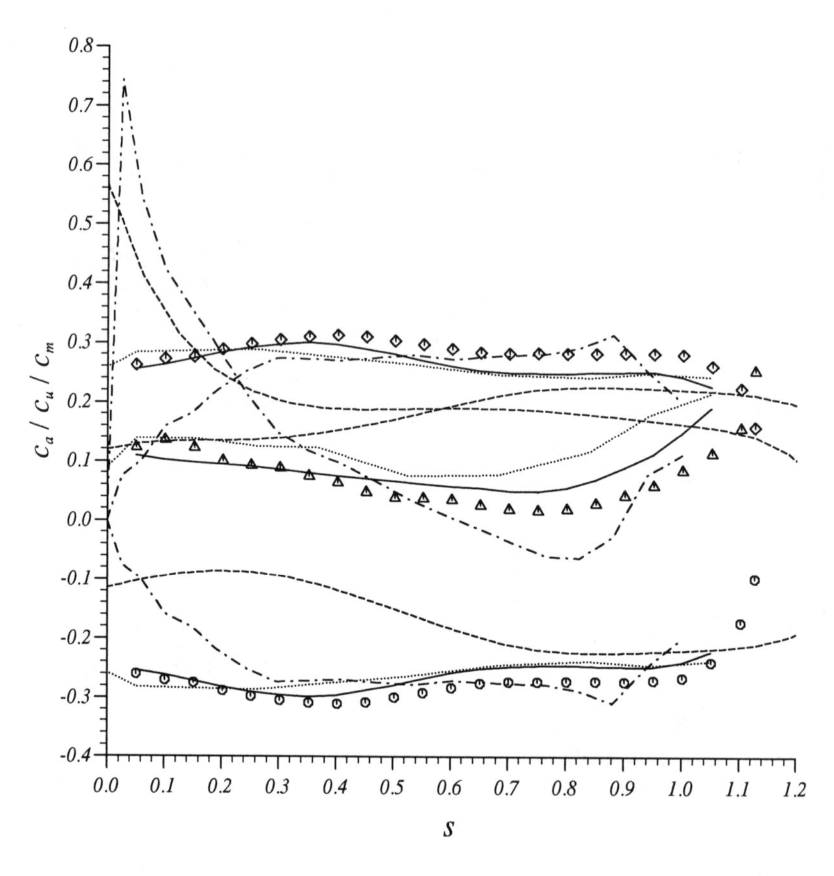

Fig. 20. Velocity components, flow angles and pressure coefficient on the runner middle axis. Comparison between the experimental data and the computed results from

—— *Arakawa/Samejima/Kubota/Suzuki {1}*
.... *Liess/Ecer {9}*
----- *Lymberopoulos/Giannakoglou/Chaviaropoulos/Papailiou {10}*
–·– *Vu/Shyy {13}*

Runner - Middle axis

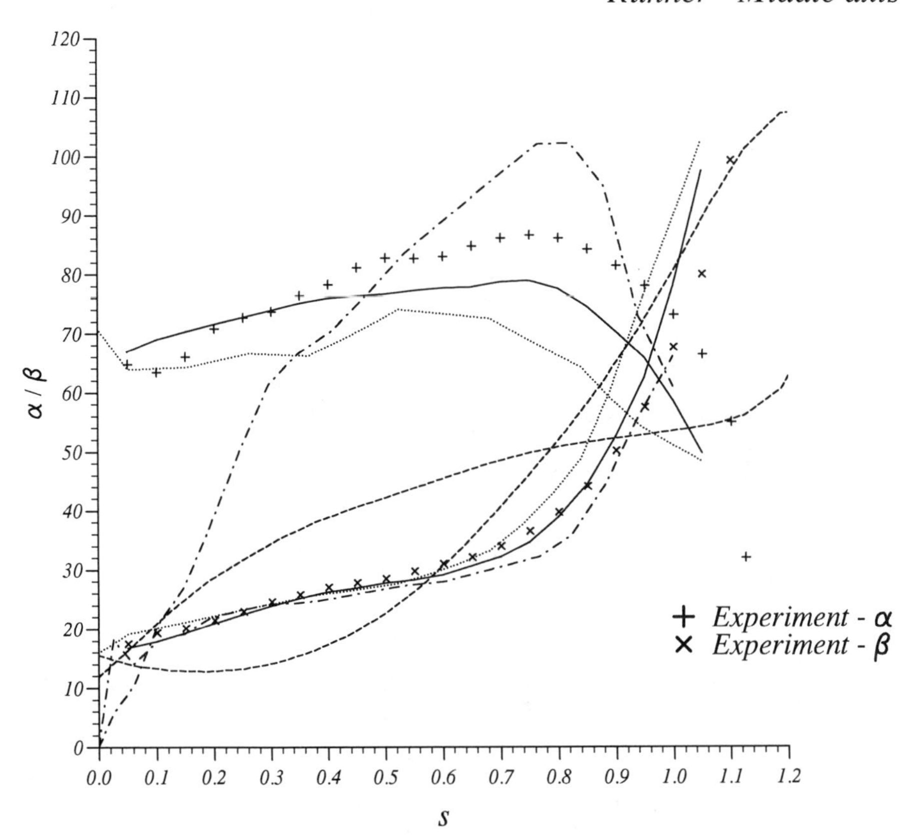

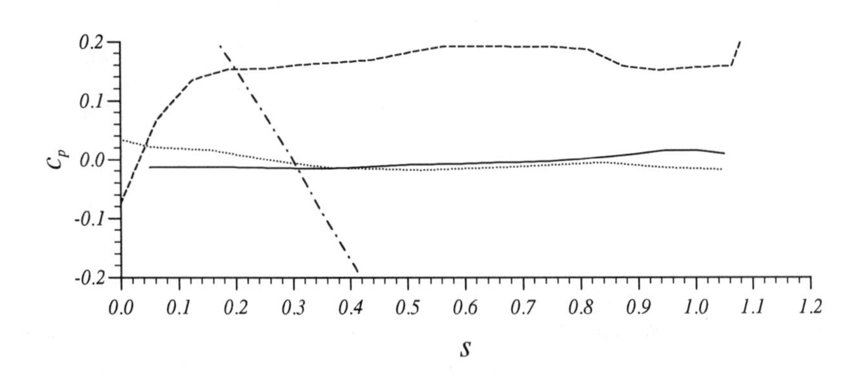

Fig. 20. (Contd.)

Runner - Outlet axis

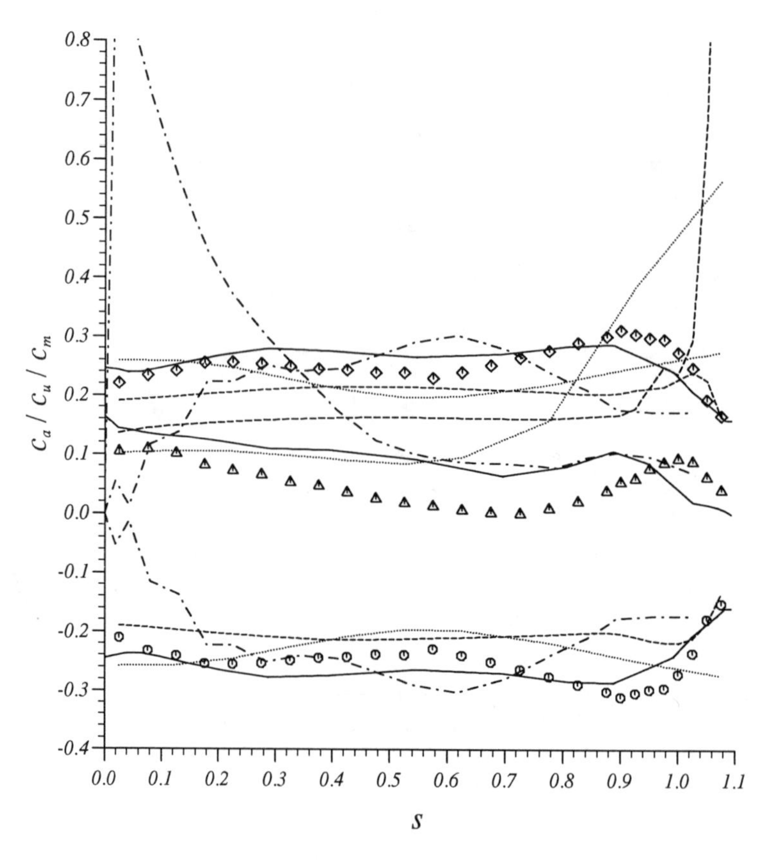

○ *Experiment* - c_a
△ *Experiment* - c_u
◇ *Experiment* - c_m

Fig. 21. Velocity components, flow angles and pressure coefficient on the runner outlet axis. Comparison between the experimental data and the computed results from

— *Arakawa/Samejima/Kubota/Suzuki {1}*
······ *Liess/Ecer {9}*
---- *Lymberopoulos/Giannakoglou/Chaviaropoulos/Papailiou {10}*
-·- *Vu/Shyy {13}*

Runner - Outlet axis

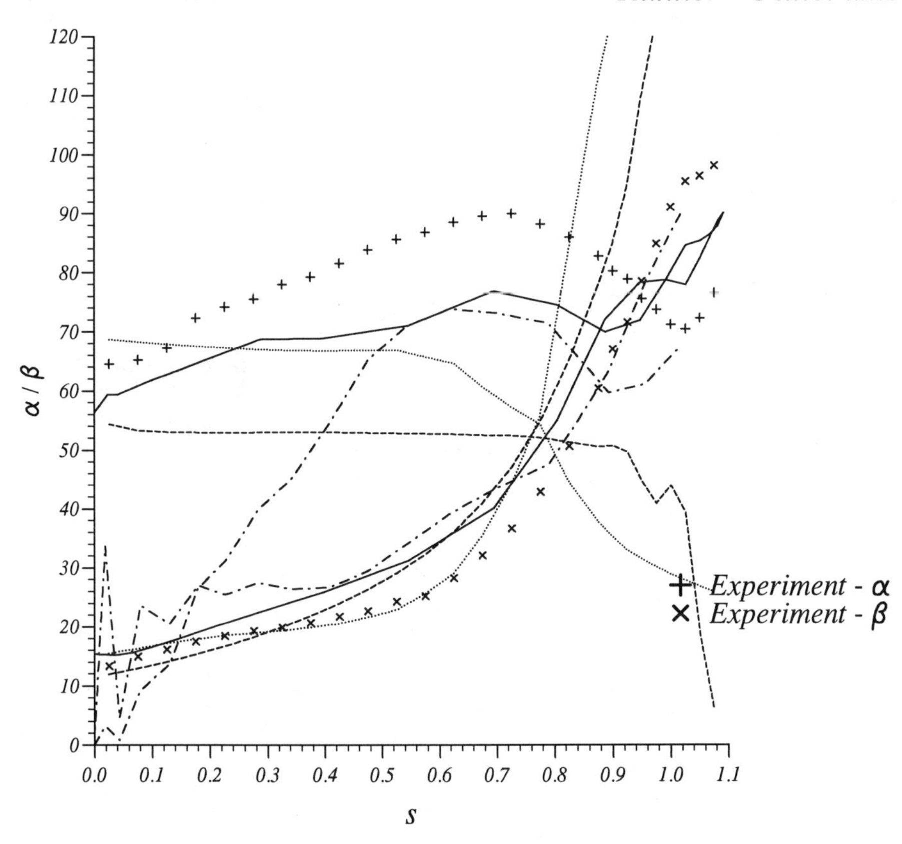

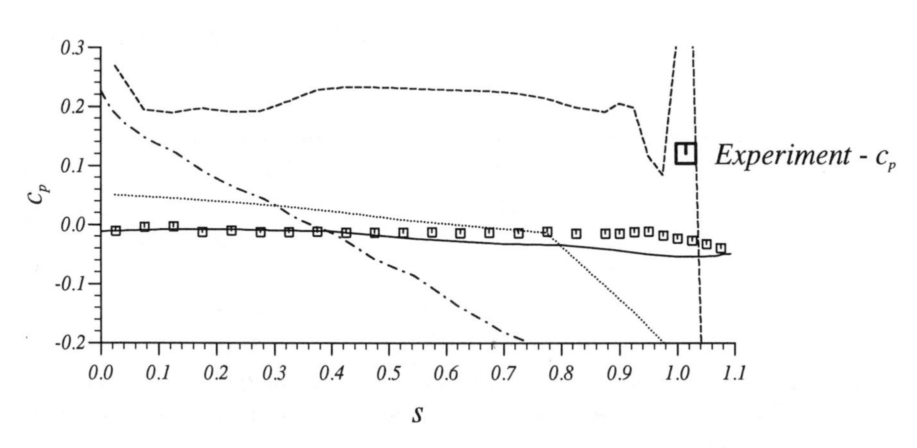

Fig. 21. (Contd.)

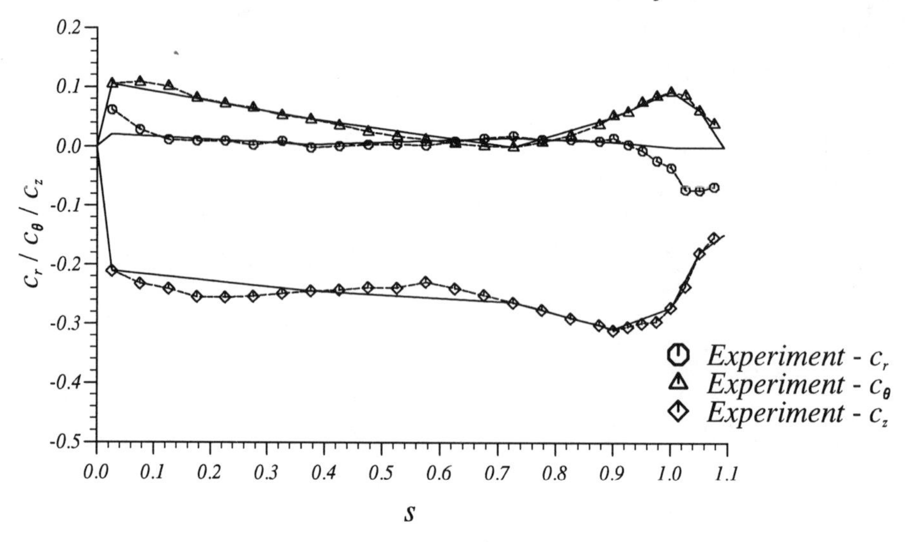

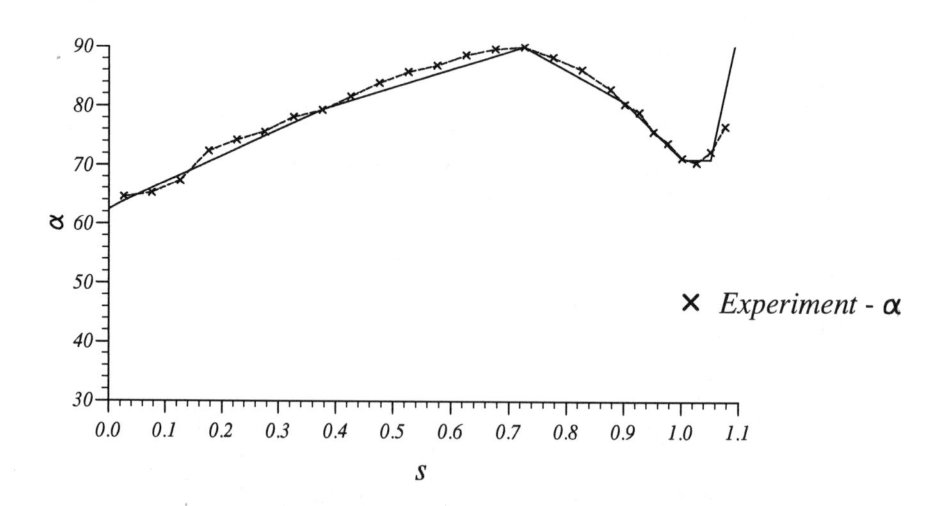

Fig. 22. Velocity components, flow angle and pressure coefficient. Comparison between the experimental data and the imposed/computed conditions at the draft tube inlet axis.

—— *Kubota/Toshigami/Suzuki*
.... *Takagi/Tanabe/Ikegawa/Mukai/Sato*
----- *Vu/Shyy*

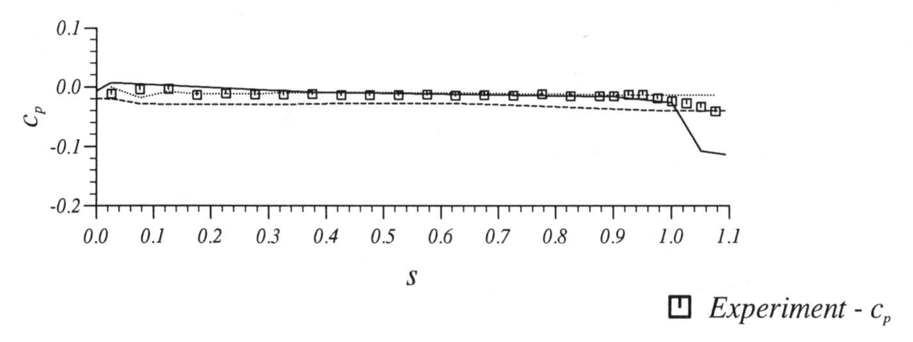

Fig. 22. (Contd.)

Draft tube axis

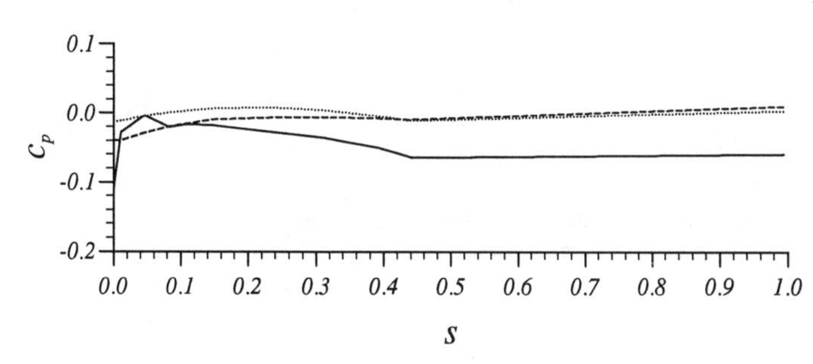

Fig. 23. Pressure distribution along the draft tube axis. Comparison between the computed results from

—— Kubota/Toshigami/Suzuki
·········· Takagi/Tanabe/Ikegawa/Mukai/Sato
----- Vu/Shyy

Draft tube - Inlet axis

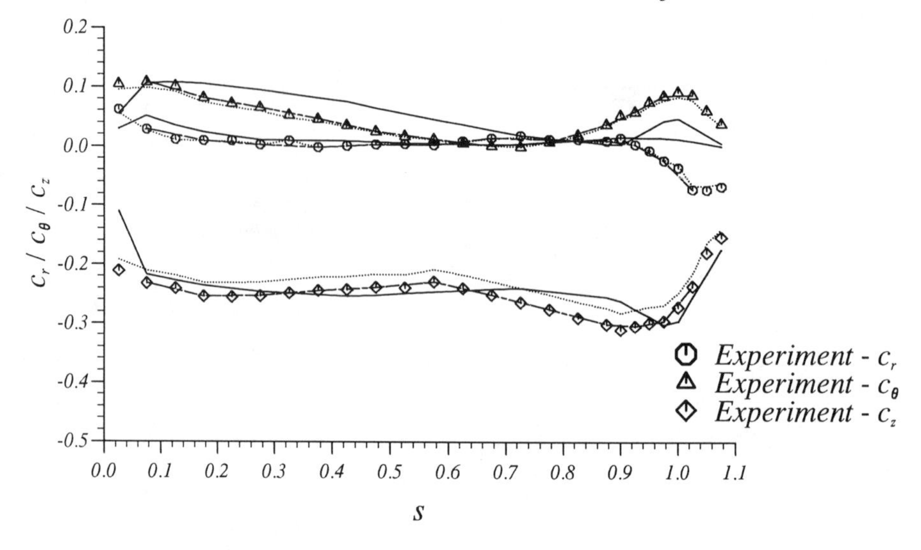

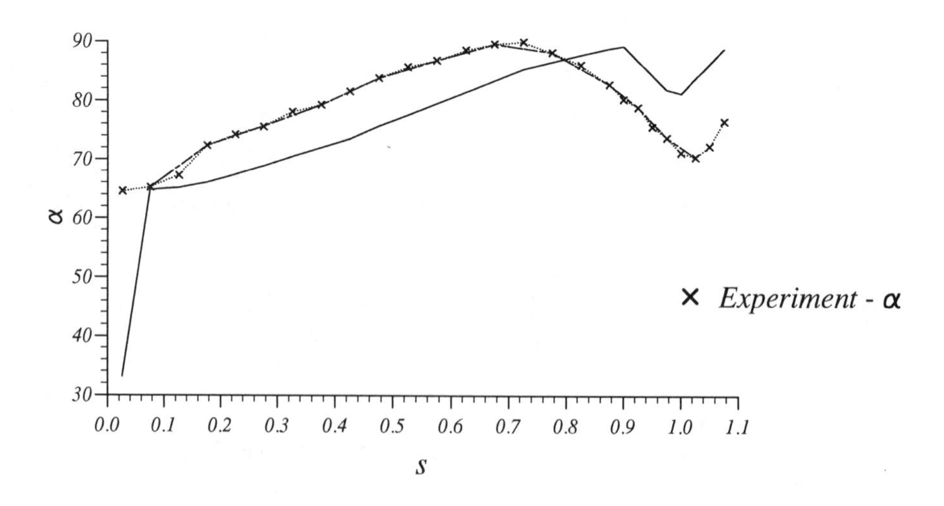

Fig. 24. Velocity components, flow angle and pressure coefficient. Comparison between the experimental data and the imposed/computed conditions at the draft tube inlet axis.

—— *Lazzaro/Riva*
········ *Liess/Ecer*
----- *Ruprecht 1*
—·— *Ruprecht 2*

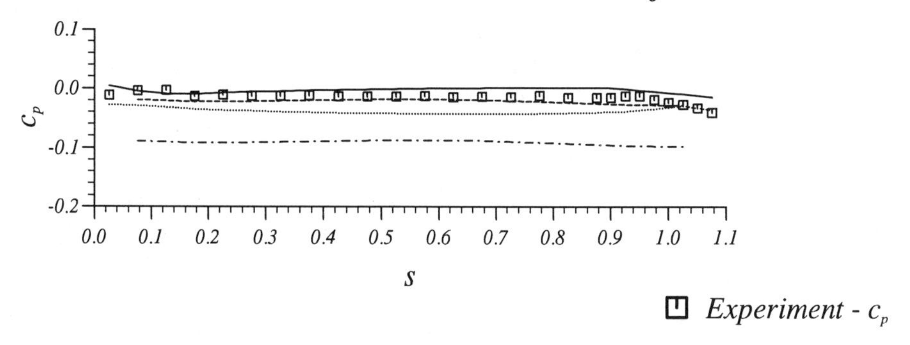

Fig. 24. (Contd.)

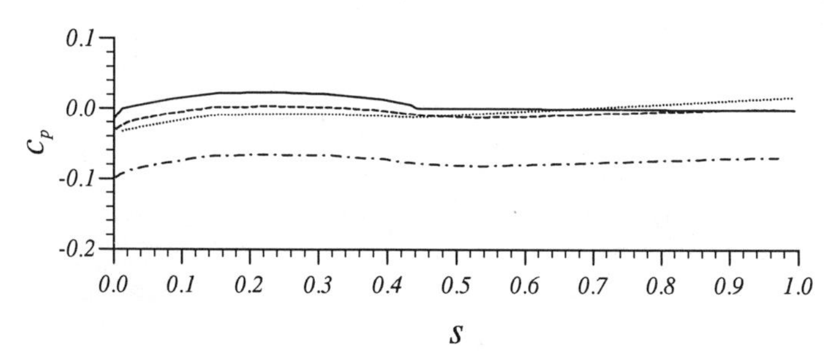

Fig. 25. Pressure distribution along the draft tube axis. Comparison between the computed results from

—— *Lazzaro/Riva*
········ *Liess/Ecer*
----- *Ruprecht 1*
-·- *Ruprecht 2*

List of Participants*

ALMQVIST Lars	Kvaerver Turbin AB P.O. Box 1005 S - 68 101 Kristinehamn, Sweden
ANDERSSON Helge	The Norwegian Institute of Technology Division of Applied Mechanics N - 7034 Trondheim, Norway
ARAKAWA Chuichi	University of Tokyo 7-3-1 Hongo, Bunkyo-ku 113 Tokyo, Japan
AVELLAN François	Swiss Federal Institute of Technology (EPFL) Hydraulic Machines and Fluid Mechanics Institute (IMHEF) Avenue de Cour 33 CH - 1007 Lausanne, Switzerland
BERGMAN Magnus	CERFACS 42, avenue Gustave Coriolis F - 31057 Toulouse Cédex, France
BILLDAL Jan-Tore	SINTEF Division of Fluid Dynamics N - 7034 Trondheim - NTH, Norway
BOTTARO Alessandro	Swiss Federal Institute of Technology (EPFL) Hydraulic Machines and Fluid Mechanics Institute (IMHEF) ME - Ecublens CH - 1015 Lausanne, Switzerland
BRATSBERG Knut	Kvaerner Brug A/S P.O. Box 3610, Gamlebyen N - 0135 Oslo 1, Norway
BREKKE Hermod	The Norwegian Institute of Technology Waterpower Laboratory Alfred Getz veg 4 N - 7034 Trondheim, Norway
DELORME Marc	Neyrpic 75, rue Général - Mangin, B.P. 75 F - 38041 Grenoble Cédex, France

* Contributor's names are in bold type.

DROTZ Alain	Swiss Federal Institute of Technology (EPFL) Hydraulic Machines and Fluid Mechanics Institute (IMHEF) ME - Ecublens CH - 1015 Lausanne, Switzerland
DULIKRAVICH George S.	Pennsylvania State University Aerospace Engineering Department 233 Hammond Building University Park PA 16802, USA
DUPONT Philippe	Swiss Federal Institute of Technology (EPFL) Hydraulic Machines and Fluid Mechanics Institute (IMHEF) Avenue de Cour 33 CH - 1007 Lausanne, Switzerland
ECER Akin	Technalysis Incorporated 7120 Waldemar Drive Indianapolis IN 46268, USA
EL GHAZZANI E. M.	Metraflu (Ecole Centrale de Lyon) 64 chemin des Mouilles F - 69134 Ecully, France
ELIASSON Peter	FFA P.O. Box 11021 S - 161 11 Bromma, Sweden
FAHRAT Mohamed	Swiss Federal Institute of Technology (EPFL) Hydraulic Machines and Fluid Mechanics Institute (IMHEF) Avenue de Cour 33 CH - 1007 Lausanne, Switzerland
FALLER Wolfgang	Sulzer - Escher Wyss Gmbh P.O. Box 13 80 D - 7980 Ravensburg, Germany
FRANCOIS Maryse	Neyrpic 75, rue Général - Mangin, B.P. 75 F - 38041 Grenoble Cédex, France
GAMBA Patrick	Swiss Federal Institute of Technology (EPFL) Hydraulic Machines and Fluid Mechanics Institute (IMHEF) ME - Ecublens CH - 1015 Lausanne, Switzerland
GINDROZ Bernard	Swiss Federal Institute of Technology (EPFL) Hydraulic Machines and Fluid Mechanics Institute (IMHEF) Avenue de Cour 33 CH - 1007 Lausanne, Switzerland

GOEDE Eberhard	Sulzer Escher Wyss P.O. Box CH - 8023 Zürich, Switzerland
HAAS Holger	Technische Universität München Laboratorium für hydraulische Maschinen und Anlagen Arcisstrasse 21, Postfach 20 24 20 D - 8000 München 2, Germany
HENRY Pierre	Swiss Federal Institute of Technology (EPFL) Hydraulic Machines and Fluid Mechanics Institute (IMHEF) Avenue de Cour 33 CH - 1007 Lausanne, Switzerland
HUSSAIN Mahmood	Swiss Federal Institute of Technology (EPFL) Hydraulic Machines and Fluid Mechanics Institute (IMHEF) Avenue de Cour 33 CH - 1007 Lausanne, Switzerland
JACOBSEN Oeivind	SINTEF Division of Fluid Dynamics N - 7034 Trondheim - NTH, Norway
KECK Helmut	Sulzer Escher Wyss P.O. Box CH - 8023 Zürich, Switzerland
KUBOTA Takashi	Fuji Electric Co. Ltd. 1-1 Tanabeshinden, Kawasaki-ku 210 Kawasaki, Japan
Current adress :	Kanagawa University Faculty of Mechanical Engineering 3-27-1 Rokkakubashi, Kanagawa-ku 221 Yokohama, Japan
LAZZARO Bruno	Riva Hydroart S.p.A. Via Stendhal 34 I - 20144 Milano, Italy
LEITNER Josef	Voest - Alpine Maschinenbau Gmbh Lunzerstrasse 64, Postfach 6 A - 4031 Linz, Austria
LIESS Christian	J.M. Voith GmbH Abt. thn Postfach 1940 D - 7920 Heidenheim, Germany
MAKINO Mitsuhiro	Fujitsu Ltd. 17-25 Shinkamata 1 Chome, Ohta-ku 144 Tokyo, Japan

MUELLER Michel
Ateliers de Constructions Mécaniques de Vevey S.A.
Rue des 2 Gares
CH - 1800 Vevey, Switzerland

NEURY Claude
Ateliers de Constructions Mécaniques de Vevey S.A.
Rue des 2 Gares
CH - 1800 Vevey, Switzerland

PAPAILIOU Kyriakos D.
National Technical University of Athens
Dept. of Mechanical Engineering
Lab. of Thermal Turbomachines
P.O. Box 64069
157 10 Athens, Greece

PARKINSON Etienne
Swiss Federal Institute of Technology (EPFL)
Hydraulic Machines and Fluid Mechanics Institute (IMHEF)
Avenue de Cour 33
CH - 1007 Lausanne, Switzerland

RICHTER Roland
Swiss Federal Institute of Technology (EPFL)
Hydraulic Machines and Fluid Mechanics Institute (IMHEF)
ME - Ecublens
CH - 1015 Lausanne, Switzerland

RIVA Pierluigi
Riva Hydroart S.p.A.
Via Stendhal 34
I - 20144 Milano, Italy

RUPRECHT Albert
Universität Stuttgart
Institut für Hydraulische Strömungsmaschinen
Pfaffenwaldring 10
D - 7000 Stuttgart 80, Germany

RYHMING Inge L.
Swiss Federal Institute of Technology (EPFL)
Hydraulic Machines and Fluid Mechanics Institute (IMHEF)
ME - Ecublens
CH - 1015 Lausanne, Switzerland

SANTAL Olivier
Swiss Federal Institute of Technology (EPFL)
Hydraulic Machines and Fluid Mechanics Institute (IMHEF)
Avenue de Cour 33
CH - 1007 Lausanne, Switzerland

SCHILLING Rudolf
Technische Universität München
Laboratorium für hydraulische Maschinen und Anlagen
Arcisstrasse 21, Postfach 20 24 20
D - 8000 München 2, Germany

SOTTAS Gabriel
Swiss Federal Institute of Technology (EPFL)
Hydraulic Machines and Fluid Mechanics Institute (IMHEF)
ME - Ecublens
CH - 1015 Lausanne, Switzerland

STRINNING Per	ITT - Flygt AB Hydraulic Design Group RD&E Box 1309 S - 17125 Solna, Sweden
SUZUKI Ryoji	Fuji Electric Co. Ltd. 1-1 Tanabeshinden, Kawasaki-ku 210 Kawasaki, Japan
TAKAGI Takeo	Hitachi Ltd. Mechanical Engineering Research Laboratory 502 Kandatu-machi, Tsuchiura-shi 300 Ibaraki, Japan
TANAKA Hiroshi	Toshiba Corporation Hydraulic Machinery 2-4 Suehiro-cho, Tsurumi-ku 230 Yokohama, Japan
TOSHIGAMI Kohji	Fuji Electric Co. Ltd. 1-1 Tanabeshinden, Kawasaki-ku 210 Kawasaki, Japan
VERRY Alain	Electricité de France (EDF) Direction des études et recherches Département Machines 6,quai Watier, B.P. 49 F - 78401 Chatou Cédex, France
VU Thi C.	Dominion Engineering Works/GE CANADA 795 First Avenue Lachine H8S 2S8 Québec, Canada
WATZELT Christian	Technische Universität München Laboratorium für hydraulische Maschinen und Anlagen Arcisstrasse 21, Postfach 20 24 20 D - 8000 München 2, Germany
WINKLER Stephan	Voest - Alpine Maschinenbau Gmbh Lunzerstrasse 64, Postfach 6 A - 4031 Linz, Austria

Contributors represented by another participant

CHAVIAROPOULOS P.
National Technical University of Athens
Dept. of Mechanical Engineering
Lab. of Thermal Turbomachines
P.O. Box 64069
157 10 Athens, Grcccc
(represented by Papailiou K.D.)

GIANNAKOGLOU K.
National Technical University of Athens
Dept. of Mechanical Engineering
Lab. of Thermal Turbomachines
P.O. Box 64069
157 10 Athens, Greece
(represented by Papailiou K.D.)

GRIMBERT Ilana
Electricité de France (EDF)
Direction des études et recherches
Département Machines
6,quai Watier, B.P. 49
F - 78401 Chatou Cédex, France
(represented by Verry A.)

IKEGAWA M.
Hitachi Ltd.
Mechanical Engineering Research Laboratory
502 Kandatu-machi, Tsuchiura-shi
300 Ibaraki, Japan
(represented by Takagi T.)

KOBAYASHI T.
University of Tokyo
Institute of Industrial Science
7-22-1 Roppongi, Minato-ku
Tokyo, Japan
(represented by Tanaka H.)

LYMBEROPOULOS N.
National Technical University of Athens
Dept. of Mechanical Engineering
Lab. of Thermal Turbomachines
P.O. Box 64069
157 10 Athens, Greece
(represented by Papailiou K.D.)

MUKAI H.
Hitachi Ltd.
Mechanical Engineering Research Laboratory
502 Kandatu-machi, Tsuchiura-shi
300 Ibaraki, Japan
(represented by Takagi T.)

NAGAFUJI Tomotatsu

Toshiba Corporation
Hydraulic Research Laboratory
20-1 Kansei-cho, Tsurumi-ku
230 Yokohama, Japan
(represented by Tanaka H.)

SAMEJIMA M.

University of Tokyo
7-3-1 Hongo, Bunkyo-ku
113 Tokyo, Japan
(represented by Arakawa C.)

SATO J.

Hitachi Ltd.
Hitachi Works
3-1-1 Saiwai-chou, Hitachi-shi
317 Ibaraki, Japan
(represented by Takagi T.)

SHYY W.

University of Florida
Department of Aerospace Engineering,
Mechanics and Engineering Science
231 Aerospace Bldg
Gainesville
Florida 32611, USA
(represented by Vu T.C.)

SUZUKI T.

Toshiba Corporation
Hydraulic Research Laboratory
20-1 Kansei-cho, Tsurumi-ku
230 Yokohama, Japan
(represented by Tanaka H.)

TANABE T.

Hitachi Ltd.
Mechanical Engineering Research Laboratory
502 Kandatu-machi, Tsuchiura-shi
300 Ibaraki, Japan
(represented by Takagi T.)

TANIGUCHI N.

University of Tokyo
Institute of Industrial Science
7-22-1 Roppongi, Minato-ku
Tokyo, Japan
(represented by *Tanaka H.*)

Notes on Numerical Fluid Mechanics (NNFM) Volume 39

Volumes 1 to 7, 9 to 11, 15, 16, 19 and 21 are out of print.

Addresses of the Editors of the Series "Notes on Numerical Fluid Mechanics"

Brief Instruction for Authors

Manuscripts should have well over 100 pages. As they will be reproduced photomechanically they should be typed with utmost care on special stationary which will be supplied on request.
In print, the size will be reduced linearly to approximately 75 per cent. Figures and diagrams should be lettered accordingly so as to produce letters not smaller than 2 mm in print. The same is valid for handwritten formulae. Manuscripts (in English) or proposals should be sent to the general editor, Prof. Dr. E. H. Hirschel, Herzog-Heinrich-Weg 6, D-8011 Zorneding.